METHODS IN MOLECULAR BIOLOGY

Series Editor
John M. Walker
School of Life and Medical Sciences
University of Hertfordshire
Hatfield, Hertfordshire, UK

For further volumes:
http://www.springer.com/series/7651

For over 35 years, biological scientists have come to rely on the research protocols and methodologies in the critically acclaimed *Methods in Molecular Biology* series. The series was the first to introduce the step-by-step protocols approach that has become the standard in all biomedical protocol publishing. Each protocol is provided in readily-reproducible step-by step fashion, opening with an introductory overview, a list of the materials and reagents needed to complete the experiment, and followed by a detailed procedure that is supported with a helpful notes section offering tips and tricks of the trade as well as troubleshooting advice. These hallmark features were introduced by series editor Dr. John Walker and constitute the key ingredient in each and every volume of the *Methods in Molecular Biology* series. Tested and trusted, comprehensive and reliable, all protocols from the series are indexed in PubMed.

Ice Binding Proteins

Methods and Protocols

Edited by

Ran Drori

Dept. of Chemistry and Biochemistry, Yeshiva University, New York, NY, USA

Corey A. Stevens

Department of Biological Engineering, Massachusetts Institute of Technology, Cambridge, MA, USA

Humana Press

Editors
Ran Drori
Dept. of Chemistry and Biochemistry
Yeshiva University
New York, NY, USA

Corey A. Stevens
Department of Biological Engineering
Massachusetts Institute of Technology
Cambridge, MA, USA

ISSN 1064-3745 ISSN 1940-6029 (electronic)
Methods in Molecular Biology
ISBN 978-1-0716-3502-5 ISBN 978-1-0716-3503-2 (eBook)
https://doi.org/10.1007/978-1-0716-3503-2

This Humana imprint is published by the registered company Springer Science+Business Media, LLC, part of Springer Nature.
The registered company address is: 1 New York Plaza, New York, NY 10004, U.S.A.

Paper in this product is recyclable.

Preface

The crystallization and subsequent growth of ice is lethal for most organisms, imposing a selective pressure on organisms that are routinely exposed to ice or freezing conditions. As a result, evolution has produced proteins that inhibit ice growth in a non-colligative manner by adsorbing to the ice-water interface. These proteins, termed ice-binding proteins (IBPs), have been found in all kingdoms of life. Organisms have adapted IBPs to match their environment. For instance, freezing-intolerant organisms, like fish and insects, use IBPs as antifreeze proteins (AFPs), to prevent the growth of seed ice crystals. On the other hand, freeze-tolerant organisms, like plants, use them to inhibit ice recrystallization. Further, some aquatic microorganisms use IBPs as adhesins to attach to ice crystals to change their location or better colonize their environmental niche. Ice binding is a unique ability; the vast majority of molecules and ions are excluded from ice and pushed away by an ice front while it grows. To achieve this feature, these molecules selectively recognize ice in a large excess of water, a remarkable biomolecular recognition challenge.

IBPs bind to the surface of ice crystals, which form during the freezing process and is termed ice nucleation. Freezing, or ice nucleation, occurs via a two-step process: (1) the formation of a stable ice embryo, or nucleus, and (2) the subsequent growth of the ice embryo into an ice crystal. In the absence of a freezing catalyst, water will remain liquid until −38 °C. However, freezing in the environment occurs at much warmer subzero temperatures. This freezing is trigged by a catalyst, termed a nucleator. There are many natural ice nucleating molecules including mineral dust, soot, and biological particles. However, the most efficient ice nucleation molecule is a membrane associated protein found in a plant pathogenic bacterium *Pseudomonas synrigae.* This protein, known as an ice-nucleating protein, or INP, catalyzes ice formation by providing an active site that organizes water molecules into an ice embryo and causes freezing near −2 °C. The *P. synrigae* INP is a particularly effective nucleator because individual proteins self-assemble into a large aggregate forming an even larger ice embryo. This unique function has been leveraged and commercialized, as attenuated *P. synrigae* expressing the INP is used to make snow at ski resorts (Snomax®). Other biological INPs have been found in a wide variety of organisms including fungi and other bacteria. The ability to initiate ice formation will impact and revolutionize energy consumption, environmental ice loss, as well as the storage of food, tissue, and organs.

Studying IBPs in the laboratory is often a challenging task, as specialized tools are required to extract and purify them, and the methods needed for quantifying activity are unique and home-developed. This point highlights the importance of this book, which specifies in detail how to obtain and characterize these IBPs. This volume of *Methods in Molecular Biology* is divided into three sections. Part I describes methods of identifying, isolating, and purifying ice-binding proteins. Part II outlines various techniques to quantify, measure, and characterize both IBPs and INPs. Part III describes ways to model, chemically modify, and synthesize mimics of IBPs, as well as applications involving these proteins. It is our hope that this detailed collection of experimental protocols will help new researchers, break-the-ice, and enter this exciting field, while also supporting established researchers broaden the scope of their investigations.

New York City, NY, USA **Ran Drori**
Cambridge, MA, USA **Corey A. Stevens**

Contents

Part III Chemical Modifications and Synthesis of IBP Mimics, MD Simulations and Evolution of IBPs

Contributors

CHEN ADAR • *Institute of Biochemistry, Food Science and Nutrition, The Robert H. Smith Faculty of Agriculture, Food and Environment, The Hebrew University of Jerusalem, Rehovot, Israel*

MAYA BAR DOLEV • *Institute of Biochemistry, Food Science and Nutrition, The Robert H. Smith Faculty of Agriculture, Food and Environment, The Hebrew University of Jerusalem, Rehovot, Israel; Faculty of Biotechnology and Food Engineering, Technion - Israel Institute of Technology, Haifa, Israel*

ISAIAH C. H. BOX • *Department of Zoology, University of British Columbia, Vancouver, BC, Canada*

IDO BRASLAVSKY • *Institute of Biochemistry, Food Science and Nutrition, The Robert H. Smith Faculty of Agriculture, Food and Environment, The Hebrew University of Jerusalem, Rehovot, Israel*

ELIZABETH A. DELESKY • *Department of Civil, Environmental, and Architectural Engineering, University of Colorado Boulder, Boulder, CO, USA*

CHRISTOPHE DETREMBLEUR • *Center for Education and Research on Macromolecules (CERM) CESAM Research Unit, Department of Chemistry, University of Liège, Liège, Belgium*

ARTHUR L. DEVRIES • *Department of Evolution, Behavior and Ecology, University of Illinois, Urbana Champaign, Urbana, IL, USA*

RAN DRORI • *Department of Chemistry and Biochemistry, Yeshiva University, New York, NY, USA*

ROSEMARY J. EUFEMIO • *Biomolecular Sciences Graduate Program, Boise State University, Boise, ID, USA*

ARPA HUDAIT • *Department of Chemistry, Chicago Center for Theoretical Chemistry, The University of Chicago, Chicago, IL, USA*

HIDEMASA KONDO • *Bioproduction Research Institute, National Institute of Advanced Industrial Science and Technology (AIST), Sapporo, Japan; Graduate School of Life Science, Hokkaido University, Sapporo, Japan*

MICHELLE LIDDY • *Department of Anatomy, School of Biomedical Science, Dunedin, New Zealand*

APARNA J. LOBO • *Department of Chemical and Biological Engineering, University of Colorado Boulder, Boulder, CO, USA*

CRAIG J. MARSHALL • *Department of Biochemistry, School of Biomedical Science, Dunedin, New Zealand*

KATIE E. MARSHALL • *Department of Zoology, University of British Columbia, Vancouver, BC, Canada*

KONRAD MEISTER • *Biomolecular Sciences Graduate Program, Boise State University, Boise, ID, USA; Max Planck Institute for Polymer Research, Mainz, Germany; Department of Chemistry and Biochemistry, Boise State University, Boise, ID, USA*

JESSICA MORRIS • *Department of Biochemistry, School of Biomedical Science, Dunedin, New Zealand*

MICHIHIRO MURAOKA • *Energy Process Research Institute, National Institute of Advanced Industrial Science and Technology (AIST), Tsukuba, Ibaraki, Japan*

YASUSHI OHYAMA • *Bioproduction Research Institute, National Institute of Advanced Industrial Science and Technology (AIST), Sapporo, Japan*

NITSAN PARIENTE • *Institute of Biochemistry, Food Science and Nutrition, The Robert H. Smith Faculty of Agriculture, Food and Environment, The Hebrew University of Jerusalem, Rehovot, Israel*

ANIKA T. RAHMAN • *Department of Laboratory Medicine and Pathology, Faculty of Medicine and Dentistry, University of Alberta, Edmonton, AB, Canada*

RALPH SCHWIDETZKY • *Max Planck Institute for Polymer Research, Mainz, Germany*

VERA SIROTINSKAYA • *Institute of Biochemistry, Food Science and Nutrition, The Robert H. Smith Faculty of Agriculture, Food and Environment, The Hebrew University of Jerusalem, Rehovot, Israel*

CHRISTIAN C. M. SPRONCKEN • *Department of Chemical Engineering and Chemistry & Institute for Complex Molecular Systems, Eindhoven University of Technology, Eindhoven, the Netherlands*

WIL V. SRUBAR III • *Department of Civil, Environmental, and Architectural Engineering, University of Colorado Boulder, Boulder, CO, USA*

COREY A. STEVENS • *Department of Biological Engineering, Massachusetts Institute of Technology, Cambridge, MA, USA*

SAKAE TSUDA • *Bioproduction Research Institute, National Institute of Advanced Industrial Science and Technology (AIST), Sapporo, Japan; Graduate School of Life Science, Hokkaido University, Sapporo, Japan; OPERANDO Open Innovation Laboratory, National Institute of Advanced Industrial Science and Technology (AIST), Tsukuba, Japan*

TYLER D. R. VANCE • *Department of Laboratory Medicine and Pathobiology, Temerty Faculty of Medicine, University of Toronto, Toronto, ON, Canada*

KARIN R. L. VAN DER BURG • *Department of Zoology, University of British Columbia, Vancouver, BC, Canada*

ILJA K. VOETS • *Department of Chemical Engineering and Chemistry & Institute for Complex Molecular Systems, Eindhoven University of Technology, Eindhoven, the Netherlands*

THOMAS F. WHALE • *Department of Chemistry, University of Warwick, Coventry, UK; School of Earth and Environment, University of Leeds, Leeds, UK*

Part I

Isolating, Purifying and Crystallizing IBPs

Chapter 1

Identifying Ice-Binding Proteins in Nature

Arthur L. DeVries

Abstract

Organisms inhabiting freezing terrestrial, polar, and alpine environments survive because they have evolved adaptations to tolerate sub-freezing temperatures. Among these adaptations are ice-binding proteins (IBPs) which in the case of fishes and some insects have antifreeze properties which allow them to avoid freezing even at their lowest environmental temperatures. Other organisms, including some insects, microorganisms, and plants, tolerate freezing and also contain IBPs. Unlike fish and insects, their antifreeze properties (hysteresis) are minimal, but most are potent ice recrystallization inhibitors (IRIs). Microbes secrete IBPs into their immediate environment where they are thought to modify ice growth in a way that ensures a liquidous habitat in the ice and also reduces ice recrystallization. With plants, IBPs are found in the small amount of apoplastic fluid associated with the extracellular spaces and show a weak hysteresis but are potent IRIs.

Techniques are described for drawing blood and hemolymph from fish and insects, respectively, in order to determine whether there is a hysteresis present (separation of the freezing and melting points) indicative of an antifreeze protein. For microbes, which secrete very small amounts of IBPs into their environment, a technique is described where their spent growth media causes the pitting of the basal plane of an ice crystal at a temperature slightly below the media freezing point. In plants, IBPs are isolated from the apoplastic fluids of the leaves by vacuum infiltration of a fluid into the extracellular spaces and then recovering the fluid by centrifugation.

The pitting of the basal plane again can be used to verify the presence of IBPs in the concentrated apoplastic fluid.

The techniques describe how to collect fluids from a variety of organisms to determine if IBPs are present using nanoliter osmometry or using the ice basal plane pitting technique.

Key words Antifreeze proteins, Ice-binding proteins, Thermal hysteresis (TH), Freeze avoidance, Freezing tolerance, Inhibition of recrystallization, Ice pitting assay, Insects, Supercooling, Nucleation, Fish

1 Introduction

Ice-binding proteins are associated with many different organisms in sub-freezing environments. In fishes in the polar oceans, the IBPs act as antifreeze agents lowering their freezing points below that of seawater (−1.9 °C) and thus avoiding freezing. Some insects

Ran Drori and Corey A. Stevens (eds.), *Ice Binding Proteins: Methods and Protocols*, Methods in Molecular Biology, vol. 2730, https://doi.org/10.1007/978-1-0716-3503-2_1,

also have IBPs that are antifreeze agents and lower their freezing points below the coldest temperature experienced in their terrestrial environment. Other insects are freeze tolerant and survive the effects of freezing and thawing. IBPs are also present in their hemolymph but at very low concentrations and tend to function as potent ice recrystallization inhibitors (IRIs). They also have macromolecules, which are ice-nucleating agents, which promote freezing at relatively high temperatures. Freezing at relatively high temperatures results in slow ice growth and in the presence of IRIs prevents the transformation of the small ice crystals into large crystals, which cause damage to the insect tissues. IBPs are also associated with microorganisms where they secrete them into their immediate environment. The consensus is that they modify ice growth in their snow or sea ice habitat so that the ice surfaces develop voids, which retain water providing the necessary aqueous environment for their survival. Higher plants that tolerate freezing secrete IBPs into their apoplastic space where freezing first occurs. They have weak antifreeze properties but are potent IRIs, which prevent the growth of large damaging ice crystals. The full complement of IBPs appears only after a lengthy period of cold acclimatization. The identification of IBPs in fishes and insects is straightforward because their blood and hemolymph are easily sampled. With plants, the IBPs in the apoplastic fluid are not directly accessible, and therefore, they have to be flushed out by vacuum infiltration of a liquid into this space and recovering it by centrifugation. The IBPs secreted by microorganisms can be recovered from their spent culture media if they can be cultured. The detailed approaches for identifying the presence of IPBs in each of the four groups are described below.

The first investigations into resistance to freezing of fishes originated in 1953 with a simple question by Dr. R.H. Backus, a marine biologist from Woods Hole Oceanographic Institution, directed to the comparative physiologist, Dr. Pere Scholander. "When arctic fishes swim about in ice-laden water at −1.7 to −1.8 °C, why don't they freeze? Do they have twice as high an osmotic concentration as ordinary fishes, or what is the story?" [1]. Temperate and tropical marine fishes have osmotic concentrations equivalent to a freezing point depression of 0.6–0.7 °C. To answer the question, Scholander and colleagues made several trips to the fjords of northern Labrador in 1955–1956 and described sculpins and fjord cod that inhabited ice-laden seawater at −1.8 °C that were immune to freezing in their natural habitat as long as they were not exposed to temperatures below the freezing point of seawater. The reported freezing points of these fishes in the winter varied between −1 and −1.2 °C [1]. Blood plasma collected from two species, a sculpin, *Myoxocephalus scorpius*, and fjord cod, *Gadus ogac*, were returned to a state-side laboratory and analyses carried out to determine the nature of the blood solutes that conferred this "antifreeze" characteristic [2]. Ion analyses as well as ones for organic solutes such as fatty acids, free amino acids, urea, and trimethylamine oxide showed that these solutes were only slightly elevated. The only unusual observation was that the non-protein nitrogen (NPN) was significantly elevated in the protein-free supernatants of plasma treated with protein-precipitating agents but only in the fjord cod. The nitrogen-containing small organic solutes accounted for only a small fraction of the NPN [2]. Curiously, even though no compound associated with the extreme resistance to freezing was identified, they termed the unknown entity an "antifreeze." This was the status of the search for an antifreeze compound in 1968 when DeVries also identified high NPN levels in the Antarctic fish *Pagothenia borchgrevinki* blood serum [3]. It was found to be due to the presence of relatively high molecular weight molecules, which were soluble in protein-precipitating agents and retained within dialysis tubing. Having determined that the NPN remained in the soluble fraction of blood serum protein precipitants, the soluble fraction was subjected to column chromatography (ion exchange and gel filtration on Sephadex G-200 columns) and was determined to be a glycoprotein composed of many size isoforms [4]. The pure glycoproteins were found to be composed of two amino acids alanine and threonine, of which the latter residue was linked to the disaccharide galactose N-acetylgalactosamine. They were also characterized with regard to molecular weight and their non-colligative depression of the "freezing point." The surprising discovery was that they lowered the freezing point (the temperature of ice growth in the presence of a seed ice crystal) well below the freezing point of seawater (−1.9 °C), but had a minimal effect on its melting point [5]. The accepted explanation for this non-colligative lowering of the freezing point that was put forward was one of adsorption to ice and inhibition of its growth [6]. All AFPs (IBPs) bind to specific ice crystal planes and divide the growth front into many small domains, which have highly curved fronts, the effect of which is the depression of the local freezing point. As indicated, the AFGPs have a small effect on the melting point with a hysteresis of approximately 0.2 °C in the blood of Antarctic fishes. The hysteresis melting point occurs because the bound AFPs impede the movement of the water molecules from the solid phase to the liquid phase [7, 8]. This melting hysteresis is also observed with endogenous ice crystals in live Antarctic fishes [8].

The serum melting and freezing points in Antarctic fishes are separated by approximately 1.5 °C and referred to as a hysteresis. This has turned out to be the hallmark of all fish antifreeze proteins and later other ice-binding proteins (IBPs) identified in insects, microbes, and plants.

Fig. 1 Notothenioid fish, *Pagothenia borchgrevinki*, in shallow water of McMurdo Sound resting on platelet ice in tunnels and crevices in the ice foot frozen to the shore. Water temperature is −1.9 °C

Fishes that inhabit shallow ice-covered freezing seawater will always possess an antifreeze because supercooling in the presence of ice is not possible as the integument (skin and especially the gills) are poor barriers to ice propagation (*see* Fig. 1) [8].

However, fishes that lack AFPs can behaviorally avoid freezing in ice-covered deep water if they avoid the surface ice. This freeze avoidance strategy is observed in deep water fjords where the water temperature is near its freezing point but ice-free. Most fishes inhabiting these ice-free waters lack antifreeze proteins and survive in a supercooled state. As long as they avoid the icy surface waters throughout their life cycle, they are safe even in this metastable state [1].

IBPs in Insects Unlike marine fishes, which experience temperatures in ice-laden water no lower than the freezing point of seawater (−1.9 °C in the polar oceans), terrestrial insects can be exposed to temperatures as low as −50 °C in the polar regions during the winter [9]. Two strategies are recognized by which they avoid the potential lethal effects of low temperatures: some avoid freezing utilizing antifreeze proteins to lower their freezing and supercooling points, while others freeze and tolerate the effects of freezing. Insects that avoid freezing harbor moderate levels of AFPs that however exhibit a large hysteresis, which does not correlate with their critical supercooling temperature (temperature at which they spontaneously freeze). The most likely reason for this disparity is that one cannot measure the hemolymph hysteresis in the presence of seed crystals less than 1 μm. With insect AFPs, hysteresis increases as the crystal size diminishes. Extrapolation of the relationship of crystal size with the freezing point indicates that at some temperature, the hemolymph freezing point would be the same as the insect supercooling point [10].

Some insects lack AFPs and avoid freezing by supercooling and do so by ridding their bodies of ice nucleators, by dehydration, by accumulating polyols, and by isolation in a dry, ice-free hibernacula [9]. With freeze-tolerant forms, biological ice nucleators (INs) are present in the hemolymph that initiate freezing at relatively high temperatures (−6 to −7 °C) resulting in slow freezing but only in the extracellular spaces [11]. Freezing at relatively high temperatures allows for osmotic equilibration to occur between the intracellular and extracellular spaces avoiding excessive supercooling of the intracellular fluid. When freezing occurs in a deeply supercooled organism, it is rapid and results in lethal intracellular freezing because there is insufficient time for water to exit the cell in response to the osmotic gradient generated by the higher vapor pressure of supercooled water relative to that of ice [12]. Freeze-tolerant insects also accumulate polyols such as glycerol and sugars such as sorbitol and trehalose, some of which equilibrate across the fluid spaces (intra- and extracellular) lowering the melting point, which provide a solvent for the high salt concentrations, which otherwise precipitate and cause cellular damage. Many also produce an IBP which exhibits only a small hysteresis (0.2–0.3 °C) but nevertheless show potent (IRI) activity [13]. This IRI activity appears to be important for preventing the damaging effects of repeated freezing and thawing by preventing growth of large crystals at the expense of small less damaging ice crystals, again, only in the extracellular spaces.

To identify insects that possess AFPs or IBPs, one has to collect them and sample their hemolymph after laboratory cold acclimation or during the winter where most often they are found under leaf litter or beneath the rotting bark of fallen tree as well as under the bark of living dormant trees. Careful examination of such insects may indicate whether or not they are frozen. Some frozen insects have external ice crystals attached to their integument, while unfrozen ones lack associated ice crystals (*see* Fig. 2).

The only sure way to determine whether AFPs are present in insects that are freeze tolerant or freeze avoidant is to sample their hemolymph and determine whether a weak or potent hysteresis factor is present. This can be done easily by removing an appendage or puncturing the exoskeleton with a small needle at the base of one of their six legs collecting the hemolymph that oozes out with a capillary tube. Of course, if an insect is too small for sampling, the only alternative is to homogenize them in an aqueous buffer, centrifuge it, and analyze the supernatant. Searching for hysteresis in the supernatant may not be straightforward as lipid micelles tend to obscure individual ice crystals when viewed in a nanoliter osmometer sample well especially if the IBPs are present in low concentrations. As with the fish blood, employing a nanoliter osmometer yields the most useful information on the freezing/melting behavior of the hemolymph and solutions of purified AFP.

Fig. 2 Frozen mountain weta, *Hemideina maori*, at −6 °C in a cavity under a rock ledge on Rock and Pillar Range in Otago, South Island, New Zealand. After approximately 2 h at room temperature, the movements of the weta appeared normal. (Photo courtesy of Hans Romlov)

IBPs Associated with Microorganisms Many microorganisms inhabiting freezing environments synthesize IBPs to mitigate the deleterious effects of ice formation in their immediate environment whether it is soil, ice, or snow. Unlike fishes and insects, which have high levels of IBPs in their extracellular spaces (blood and hemolymph spaces), there is no such space associated with microorganisms. Although most microorganisms secrete their IBPs into their immediate environment [14], there are a few where the IBPs remain attached to the outer cell wall. Being anchored to the cell wall appears to be important for attaching the microbe to ice or to other microbes such as diatoms. It is suggested that the bacterium *Marinomonas primoryensis* inhabiting a freshwater lake in Antarctica attaches itself to surface ice where it remains in the photic zone, which would be well oxygenated and rich in carbon energy sources because of the primary production by diatoms and algae [15]. By secreting IBPs into their immediate environment, microbes preserve the surrounding liquid habitat necessary for survival. In sea ice, IBPs cause distortion of ice grown in their presence leading to the formation of small brine pockets that are thought to retard the drainage of brine [16–18] providing a liquidous habitat. When saltwater ice containing IBPs is grown in a tube, it has a fibrous structure containing voids that retain the brine even when centrifuged [18]. Similar ice structuring in nature would ensure an aqueous habitat for the survival of bacteria, diatoms, and algae [17, 18]. It is difficult to relate the amount of IBP associated with microorganisms in their environments because of the absence of a confined space, but evidence suggests that the amount of IBP in their environment is very small. Thus, IBPs have been isolated from those that can be cultured by processing their cell-free spent culture media. When isolated in pure form, the specific activity of their hysteresis was found to be much lower than that of fish and insects. However, there are a few exceptions such as *Colwellia* sp., which has a hysteresis of 3.8 °C at 140 μM, while the yeast *Glaciozyma antarctica* has a hysteresis of only 0.08 °C at 200 μM [19]. Despite this variability in thermal hysteresis activity, they all exhibit potent IRI activity similar to that observed with some of the most potent insect IBP IRI activities. Interestingly, the potency of IRIs across all taxa expressing them doesn't correlate with the level of hysteresis. It does show a positive correlation though with their molecular weights with the larger molecules being more potent than the smaller ones [13]. Thus far, some of the most potent IRI proteins are those associated with the microbes and plants.

Originally, two types of IBPs were recognized in microorganisms and classified as IBP1 and IBP2. IBP1 is by far the most common, and they all share similar structures based on a similar domain composed of approximately 200 amino acids that is referred to as a domain of unknown function (3494 DUF) [19]. The less common IBP2 is found in chlorophyte alga and in a few bacteria and has a completely different structure [20].

The phylogeny of the DUF 3494 IBPs from a wide range of taxa does not resemble that of the microorganisms sampled. The explanation put forward is that organisms acquired their IBPs by horizontal gene transfer from various donners and thus a phylogeny congruent with one obtained from ribosomal genes of the host organisms would not be expected [19]. When and how these transfers occurred is not clear, and it is most likely that examining more taxa will not establish a meaningful pattern of relatedness.

Since most microorganisms secrete IPBs and if they can be cultured, the presence of IBP activity in the media can be tested using the nanoliter osmometer, which provides the most useful information. If no hysteresis, the presence of ice shaping is still indicative of ice binding, which in turn implies that IRI activity will be present. Even when no detectable hysteresis is observed, a low level of IBP can still cause an ice crystal to grow into a symmetrical hexagon or even an irregular shaped one at a temperature slightly below its melting point again indicating ice binding. If a nanoliter osmometer is not available, the presence of IBPs in water collected from their habitat or spent culture media can be detected by observing whether hexagonal pits form on the basal plane of an ice crystal at a temperature slightly below the equilibrium freezing point of the solution [14]. With this method, the only major equipment needed is a viewing chamber connected to a refrigerated bath as well as a microscope for viewing. It doesn't require the fine temperature control required for observation of single small crystals with a nanoliter-type instrument. The methodology of this approach is described below.

IBPs in Plants In contrast to fishes and some insects, plants cannot avoid freezing if cooled much below the freezing point of their aqueous fluids. Many plants tolerate freezing, and it occurs only in the fluids of the extracellular spaces such as the apoplastic fluid. Such plants are generally found in cold temperate, alpine, and polar environments, and some may be annuals but most are biannuals or perennials. As in other organisms, intracellular freezing is lethal, and thus freeze-tolerant plant adaptations are ones that preserve the integrity of their cells, particularly their cell membranes. Several IBPs have been identified in the proteins present in the apoplastic fluid of many cold acclimatized freeze-tolerant plants. In contrast to fishes and insects, in most cases, it is not possible to directly sample their apoplastic fluids. Thus, plant physiologists have developed a procedure where they can retrieve the small amount of apoplastic fluid which contains small amounts of protein from a leaf by first vacuum infusing a liquid into the apoplastic space [21]. This is done by submerging the leaf in H_2O or a dilute buffer and subjecting it to a vacuum, which removes the air in the apoplastic spaces. Releasing the vacuum allows liquid to flood the evacuated air spaces. Following the removal of the surface liquid, low-speed centrifugation for a few to several minutes expels the liquid which includes the apoplastic fluid through the stomatal openings. If the weight of the leaf is measured before infusion as well as before and after centrifugation, the initial apoplastic volume can be obtained. If this wash is reduced to a volume equal to that which was initially present in the apoplastic space, the protein concentration and any associated hysteresis can be determined and expressed per ml of apoplastic fluid. Such analyses show that the apoplastic fluid hysteresis is low relative to fish and insect fluids (0.05–0.3 °C). Purification of the individual IBPs from the fluid allows for determination of the specific activity of the IBPs, which as expected is also very low (0.1–0.3 °C) with a few exceptions [13]. It also allows for a molecular weight determination, amino acid composition, amino acid sequence, and possibly its molecular structure.

Despite the small hysteresis, the IBPs still cause alteration of ice crystal shape during freezing: a shape change from a disk to a symmetrical hexagon and rarely, if they are potent, to a hexagonal bipyramid. The shaping of the seed crystal even in the absence of hysteresis indicates that IRI activity should also be present. As it turns out, plant IBPs have potent IRI activity with the end point being in the low nM range (50 nM) similar to some bacteria [19]. Their molecular weights, however, are two to three times those of most bacteria, insects, and fishes. Interestingly, many of the plant IBPs exhibit a dual function in that they are also inhibitors of pathogen activity. They have been shown to function as chitinases, gluconases, and anti-viral thaumatin-like proteins [22]. During the summer when a non-acclimated plant expresses apoplastic proteins in response to ethylene exposure, they show both pathogen inhibitory activity and IBP activity, but when exposed to salicylic acid, they only show pathogen inhibitor activity [23]. The presence of IBPs in freeze-tolerant plants can best be observed in nature after they have first been cold acclimatized during the beginning of the winter season. Some IBPs are expressed at high levels within a few days of cold acclimation in the laboratory, whereas others only appear after several weeks of cold exposure. With temporal variation of expression and concentrations, it is difficult to assign the contributions that each type of IBP makes to the freeze tolerance of the intact plant. Nevertheless, characterization of the individual IBPs, their concentrations, and IRI potencies and interactions with cell membranes is a starting point for elucidating their role in plant freeze tolerance. Sorting out which proteins are IBPs and when they are expressed is a complicated time-consuming task but necessary to determine their role in freeze tolerance of the plants.

As with other IBPs, the presence of hysteresis in the apoplastic fluid and ice shaping can be determined with a nanoliter osmometer. More meaningful, however, is information gained by determination of their IRI potency as it appears that IRI proteins are involved in mitigating freeze-thaw damage in nature in all freeze-tolerant plants by slowing ice growth by binding to small ice crystals and preventing recrystallization. As the demonstration of IRI activity requires a low temperature-controlled cold stage, an alternative way to determine whether or not IBPs are present is the use of the ice pitting assay described for microorganisms in the previous section and if present follow-up with the IRI procedures which are more quantitative but are more complicated as discussed in another chapter. A description of the procedure for obtaining the apoplastic fluid from the extracellular spaces is described.

2 Materials

2.1 Fish Blood Collection

1. Fish anesthetic, MS-222 (tricaine methanesulfonate), Sigma.
2. 2.5 mL plastic syringe with 23-gauge needle and 1.0 mL insulin syringes with attached 27-gauge needle.
3. Small V-shaped wooden trough to position fish ventral side up.
4. Heparin solution 10,000 units/mL H_2O or fish ringers.
5. 15 mL conical plastic centrifuge tubes.
6. 1.5 mL Eppendorf tubes.

7. Fisher Brand, snapcap, polyethylene, 0.4 ml tubes, 02-681-229.
8. 10 μL Drummond capillaries.
9. Microfuge, preferably one that has a swinging bucket rotor.
10. Centrifuge capable of accepting 15 mL plastic centrifuge tubes capable of 4000× *g* like an International Clinical Centrifuge Model CL with swinging bucket rotor.

2.2 Collection of Hemolymph

1. 0.5-inch-long 30-gauge needles. Pack of 10 μL Drummond capillaries.
2. Two-sided sticky tape.
3. Eppendorf tubes, 1.5 mL.
4. Metal plate 10 cm square × 0.5 to 1 cm thick.
5. Alcohol lamp or small butane microtorch (Micro Torch, Roburn, MT-770).
6. Small low-speed mini microfuge with only on/off switch as it starts and stops quickly.
7. Ice bucket with flake ice for cooling insect.
8. Microfuge capable of 14,000× *g*.
9. Drummond capillaries, 5 μL used for pulling micropipettes.
10. PE-50 polyethylene tubing, Clay Adams.
11. Micrometer syringe, 0.2 mL, Gilmont Instruments, GS-1100.

2.3 Identification of IBPs in Microbes Using an Ice Pitting Assay

1. Glass tube, 1.3 id × 15 cm with an attached rectangular tubing extension with inside dimensions of 2.5 cm long × 1.0 cm wide and 0.6 cm deep fused to one end of the tube by a glass blowing shop using rectangular tubing R0610, VitroCom.
2. Long Pasteur pipette (22.5 cm) to attach as a handle to glass slide with super glue or 5 min Epoxy. If the sample tube is shorter than 15 cm, a shorter Pasteur pipette can be used.
3. Several precleaned microscope slides (25 × 75 × 1.0 mm) and diamond-tipped scribing tool.
4. Refrigerated viewing chamber and attached refrigerated circulator with a temperature control of +/−0.05 °C (*see* **Note 1**).
5. Polyurethane ice bucket with sufficient NaCl to lower ice/-water slurry to a temperature of ~0.1 °C below the melting point of the culture medium (*see* **Note 2**).
6. Polyurethane insulated ice bucket to grow ice crystals in a −20 °C freezer (*see* **Note 3**).
7. Dissecting microscope mounted horizontally on a stand to view the ice crystal on the slide with transmitted light.
8. Two thermometers scaled in 0.1 °C increments for both the refrigerated chamber and the ice/mixture.

2.4 Collection of Plant Apoplastic Fluid Containing IBPs

1. Side-arm vacuum flask, 100 or 200 mL.
2. Vacuum pump capable of a vacuum of 200 mmHg or a water aspirator with sufficient water pressure to generate a similar vacuum.
3. Centrifuge with swinging bucket rotor that will accept 15 mL conical centrifuge tubes such as a desktop International Clinical Centrifuge.
4. 1 mL plastic pipette tips.
5. Parafilm.
6. 15 mL plastic centrifuge tubes.
7. 1.5 mL Eppendorf tubes for storing the apoplastic wash fluid.
8. Balance to weigh to 0.1 mg.
9. Cold cabinet or cold room set at +4 °C.
10. Chest freezer −15 to −20 °C for manipulating ice crystals.

3 Methods

3.1 Collecting Blood from Fishes

1. Blood samples from fishes inhabiting freezing seawater are required for two reasons. The first is to determine whether an AFP is present by comparing the melting and freezing point to determine whether they are the same or significantly different, the latter of which is the hallmark of AFPs. The difference has been termed a thermal hysteresis, and it is correlated with the amount of blood AFP. If present, the second goal is to obtain enough blood to isolate the protein responsible for the hysteresis factor in pure form so that its composition, structure, and ice binding mechanism can be determined.
2. To obtain blood from a specimen, regardless of how they were caught, they should be anesthetized by putting them in 12 L of seawater containing 1 g of MS-222. Usually after 10 min, the immobilized fish can be bled by putting it in a V-shaped trough lined with a wet cloth or paper towel and positioned with its ventral side up. The anterior end of the V-shaped trough should be elevated approximately 2 inches to avoid blood pooling in the head region. Depending upon the size of the fish, a one inch 23 gauge needle attached to a 2.5 cc syringe is inserted into the tail region between two adjacent fin rays in the middle of the caudal fin. The caudal vein is located in the hemal canal located ventral to vertebral centrum. Care should be taken to insert the needle exactly on the mid-line at a slight angle toward the caudal fin so that it tracks parallel to the two adjacent rays. Once the needle is 1 cm in the muscle, a slight negative pressure should be applied so that when and if the vein is pierced, one will see blood immediately appearing in the

syringe. Once the needle is in the vein, it is very important to hold the syringe absolutely still with the left hand and use only the fingers and thumb of the right hand to slowly draw the blood but without excessive vacuum, thus avoiding collapse of the vein. For a 200 g fish, approximately 1.5 mL of blood can be drawn and the fish returned to fresh seawater to recover if one plans a second bleeding after a few days of recovery (*see* **Note 4**). If the needle misses the vein, retract it about 0.25 cm and angle it slightly toward either side probing again with a slight vacuum (*see* **Note 5**).

3. The blood sample should be expressed slowly into a 15 mL conical plastic centrifuge tube. If serum is being collected, the tube should be positioned horizontally so that the blood spreads out in a long narrow pool. A long narrow clot will express more serum than a solid clot at the bottom of a vertical tube. The tube should be placed on ice in a horizontal position or in a refrigerator at +4 °C and allowed to clot for at least 6 h. The reason for collecting serum as opposed to plasma is that the native ion concentrations are preserved, whereas when heparin is used to prevent clotting, some extra ions may be introduced and the sample slightly diluted. If the blood ions are not a priority, plasma can be collected by rinsing the syringe with an aqueous solution of sodium heparin containing 10,000 units/mL, which is sufficient to prevent clotting. Heparinized blood can be stored upright for a few hours before centrifugation (*see* **Note 6**).
4. For very small fish specimens (100 g or less), and especially those that have an eel-like body morph, it is best to use a 1 mL insulin syringe with a 27-gauge needle, or smaller, where the needle is embedded in the plastic tip of the syringe. With this type of syringe, almost all of the blood can be expelled because the dead volume is very small. With these very small needles, the blood must be expelled very slowly to prevent hemolysis and allowed to clot in 1.5 mL Eppendorf tubes positioned horizontally. For even smaller amounts of blood, 0.5 mL polyethylene "bullet" tubes will make serum recovery easier. If very small volumes are recovered, they can be transferred to a capillary and sealed to prevent evaporation (see capillaries in insect methods and micropipettes for recovering a sample from a sealed capillary).

For assessing AF/IBP activity, the nanoliter osmometer is the most useful for the following reasons. Sample volumes required are in the low nanoliter range, and both the melting point and freezing temperatures can be determined on the same sample. Crystal shaping can also be observed within the hysteresis gap. Depending on the concentration of the AFP, the following shapes can be observed going from low to high AFP concentrations: a symmetrical shaped hexagon, a hexagonal bipyramid of variable length, and blunt hexagonal bipyramid. The direction of the seed crystal growth at the freezing temperature (burst point) can also be observed. C-axis spicular growth has been observed with the AFGPs, while with insect AFPs, a-axes growth results in a star-shaped or floral pattern. Even with IBPs that have very low levels of hysteresis, one can observe shaping of the seed crystal at a temperature slightly below the melting point, which indicates that IRI activity will be present.

3.2 Collection of Insect Hemolymph

1. To immobilize an insect, place it in a beaker that has been cooled in an ice bucket containing crushed ice.
2. To keep the insect cold and immobilized while sampling, cool a metal plate on a bed of ice to which double-sided tape has been attached.
3. Place the insect on the tape ventral side up and use a 30-gauge needle to puncture the exoskeleton at the base of a leg where the joint exoskeleton is thin.
4. A "bead" of hemolymph should form. Holding a 10 μL capillary tube horizontally, touch the fluid, and it will be drawn in by capillary action (*see* **Note** 7).
5. If not analyzing the hemolymph immediately, using an alcohol lamp or preferably a butane microtorch, seal the end of the capillary that is free of hemolymph. If one quickly passes the torch flame over the tube a few mms from the end being sealed, it will heat the air enough so that the hemolymph will not be expelled from the open end when heating the end being sealed. Hold the capillary tip at the edge of the blue flame near the tip until sealed.

   Usually it will "bead up" slightly and appear transparent. Sealing the end should take only a few seconds. After removal from the flame, the sealed end of the capillary will cool, and the hemolymph will be drawn toward the sealed end indicating an airtight seal.
6. Next place the capillary in an Eppendorf tube with the open end up and briefly centrifuge in a mini-micro-centrifuge so that the liquid column is forced to the sealed end. A mini-microfuge is preferable because it accelerates and stops quickly with the only control being an on/off switch.
7. Before sealing the remaining open end, again briefly (less than a second) play the flame over the open end. This will heat the air inside so that while sealing the end, the expansion of the air will be insufficient to distort the end while sealing. Failure to heat the air will cause a thin-walled glass bubble to form, which is very fragile. When sealing the end, rotate the capillary at the edge of the end of the blue flame until you see it bead

up. Examination of the bead under magnification will indicate whether there is a channel to the outside or if it is completely sealed. With a little practice, this can be routinely done.

8. The sample in the sealed capillary can now be centrifuged in a benchtop microfuge at 12,000× *g* to precipitate any cells in the hemolymph. The sealed capillary will prevent evaporation and can be stored frozen preferably at −80 °C until analyzed. Rarely do these capillaries break upon freezing even at −80 °C.
9. Upon thawing, the hemolymph can be sampled using a 0.2 mL micrometer microburette with a blunt-end 23-gauge needle attached to a 20 cm length of polyethylene tubing (PE-50) with a thin pipette tip joined to the end with a short length of small flexible silicone tubing that can accommodate both the PE-50 tubing and the glass pipette. The burette, tubing, and tip are filled with light mineral oil making sure no air bubbles are present. After breaking off the empty end of the capillary tube, the fine-tipped pipette can be inserted and 1 μL withdrawn, which is sufficient to load 7 wells in a Clifton nanoliter osmometer. The capillary can be resealed if there is enough free headspace (*see* **Note 8**), and if not, expel the hemolymph on to a piece of Parafilm and immediately transfer it into a new 10 μL capillary and seal with the torch.
10. Again the nanoliter osmometer would be the instrument of choice for determining the presence of a putative IBP for the reasons given in the prior section.

3.3 Identification of IBP Activity in Microbial Spent Media by Ice Pitting of the Basal Plane of a Single Ice Crystal

1. Scribe a line parallel to the long edge of a microscope slide that is 0.8 cm from the edge and break the narrow piece off and cut it into 2.5 cm length. Glue a cut piece to the tip of Pasteur pipette (23.5 cm length) with super glue or epoxy and check to make sure this ice crystal holder fits into the rectangular end of the sample tube. Fabricate three or four holders in advance as the piece of microscope slide can easily break off. Also allow a couple of hours to make sure the bonding material is cured.
2. Set the temperature of the refrigerated circulator so that the viewing chamber is 0.1 °C below the freezing point of the native culture media.
3. Add 0.5 mL of the spent microbial culture media minus the cells to the modified test tube without the ice crystal holder and place in the ice bucket which has been adjusted to the freezing point of the native culture media by adding salt and stirring (*see* **Note 9**).
4. Place an ice crystal holder in the −20 °C freezer (*see* **Note 10**).
5. Retrieve an ice crystal from the jar of harvested crystals with a clawed forceps. Put in a rectangular cover of a pipette tip box and trim it with a warm scalpel blade to ~0.25 wide by 0.5 cm

long preserving the tip of the crystal. With the clawed forceps, transfer it to the slide surface of the precooled holder. Lift the holder with the crystal and warm the back side of the slide with your finger for 1 or 2 s. This will create a thin layer of water between the ice crystal and the glass, and it will be securely attached to the glass upon refreezing. Return to the −20 °C freezer for 2 or 3 min to ensure the crystal is firmly attached to the glass.

6. Transfer the holder with attached ice crystal to the precooled spent culture media in the ice bucket, which should be at a temperature approximately 0.05 °C below the freezing point of the native culture media.

7. After 5 min of equilibration, transfer the tube to the viewing chamber and view the ice crystal with the dissecting microscope using transmitted light. If the refrigerated viewing chamber is too cold and ice growth occurs in the media surrounding the crystal and holder, raise the temperature slightly. If the crystal face is overgrown, remove the holder and hold your finger near the exposed ice face to melt it back so that it is again a nominally flat surface. Too much melting will detach the crystal; thus, one has to be extremely careful (*see* Fig. 3).

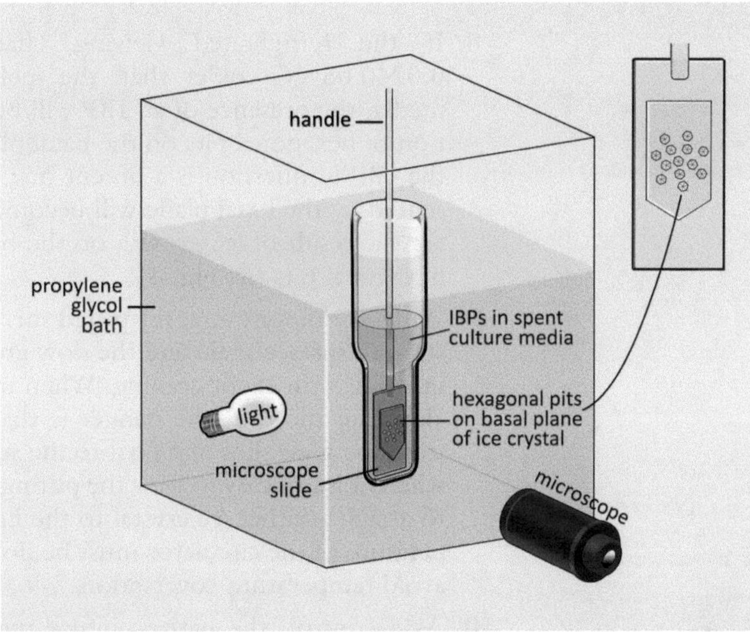

Fig. 3 Schematic of modified test tube with ice crystal frozen to a piece of a microscope slide attached to a handle. At a temperature slightly below the freezing point of an IBP solution, hexagonal pits form on the basal plane. The basal plane area slowly thickens with the formation of hexagonal pits in the newly formed ice

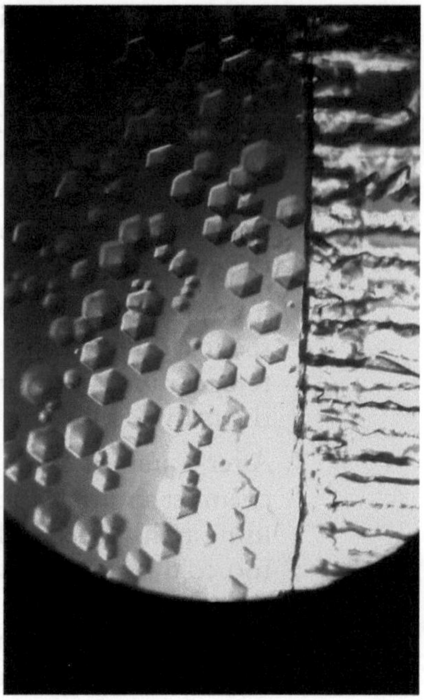

Fig. 4 Picture of hexagonal pits formed on the basal plane of an ice crystal in a solution of ice-binding protein associated with the diatom *Nitzschia stellata*. (Photo courtesy of J.A. Raymond)

8. If the refrigerated viewing chamber is approximately 0.02–0.05 °C colder than the melting point of the spent media, the presence of an IBP will be indicated by the formation of hexagonal pits on the basal plane within 20–30 min. If the IBP in question is a potent one, the entire surface represented by the basal plane will become densely pitted. The pits are the result of ice growth on the basal plane leaving behind hexagonal pits (*see* Fig. 4).
9. With careful warming, the pitted surface can be melted to a flat smooth surface again and the slow growth of the pits observed in detail with slight cooling. When melting the basal plane to eliminate the pits, the danger is that the crystal may detach from the glass slide and float to the surface and rest at an angle where it is not easy to view the pitting. If this happens, it is best to attach another ice crystal to the holder. Adjusting the temperature of the circulator must be done in small increments to avoid temperature "overshoots."
10. As a control, the native culture media should be tested to ensure that there is no pitting due to its constituents.
11. The pitting assay is qualitative, and its sensitivity at very low concentrations has not been explored.

3.4 Collecting Apoplastic Fluid by Leaf Infusion and Centrifugation

1. This description is generalized and uses a small dicot leaf that doesn't require reduction in size as an example. It also applies to small monocot leaves, which because of their length required shortening. They are usually cut 4 cm from the tip.
2. Weigh the leaf to the nearest 0.1 mg after removing the petiole.
3. Rinse the leaf in distilled water to remove associated bacteria as well as any fluid from the cut surface if the leaf has been cut. Dry the leaf by blotting on a Kimwipe.
4. Submerge the leaf in the vacuum flask containing water or a dilute buffer solution and apply vacuum for 5 min with a vacuum pump, which will remove the air from the apoplastic space (*see* **Note 11**).
5. Release the vacuum, which will allow water or buffer to fill the evacuated air spaces. Again apply vacuum for another 5 min to remove any remaining air and release it.
6. Remove the leaf and blot dry on a Kimwipe and weigh again and record.
7. Cut the end of a plastic 1 mL pipette tip so that the total length is 4.5 cm.
8. Position the leaf on a piece of Parafilm normal to the long axis (10 cm) of the same width as the pipette tip is long.
9. Place the pipette tip with the cut end in the same direction as leaf tip and gently roll the leaf and Parafilm around the pipette. After the leaf is sandwiched between the Parafilm and the pipette tip, wrap a narrow strip of Parafilm around the middle of the wrap so that it can be handled without unwinding. The narrow strip can be stretched and overlapped so that it is secure.
10. Insert the leaf wrap with the pipette tip down into a 15 mL centrifuge tube and cap.
11. Centrifuge in a cold room at +4 °C at 1200× *g* for 5 min. A clear colorless fluid should accumulate in the end of the tube, which is the apoplastic wash. A slightly green color would indicate the presence of chlorophyll either from cells of the cut end of the leaf or from ruptured cells. The centrifugal force should be great enough to recover all the apoplastic fluid, but not so great that it causes rupture of the cells. Finding the optimum centrifugation force may involve some trial and error.
12. After centrifugation, weigh the leaf again. From the weight prior to infusion, post infusion, and post centrifugation, the initial volume that occupied the apoplastic space can be determined. Also from the total volume, the concentration of the proteins can be determined using the Bio-Rad protein assay and related back to the original concentration in the apoplastic fluid.

13. To determine if there is cellular rupture, the conductivity of the wash can be measured with a micro-conductivity cell as the native apoplastic fluid is essentially free of ions. Conductivity close to that of the infused fluid would indicate no damage to the cells from infusion and centrifugation.
14. The length of time for infusing the leaves and the centrifugation force necessary to recover it will depend upon the type of plant. Some require much longer times for infusion and higher g forces to expel the liquid. In this procedure, whether the stomata are open or closed will also affect the efficiency of the infiltration and recovery. The adjustment of the g forces necessary to expel the apoplastic wash may involve trial and error as too much force will cause cell rupture, while insufficient force will result in low recovery of the wash (*see* **Note 12**).
15. To determine whether IBPs are present, the nanoliter is the method of choice. Even if no hysteresis is detected, crystal shaping would be indicative of an IBP, which may have a potent IRI activity. IRI activity analyses should be carried out to determine the potency of the IBPs. The basal plane pitting assay discussed above can also be used to identify the presence of IBPs in the event that a nanoliter osmometer is not available.

4 Notes

1. A simple usable refrigerated viewing chamber can be constructed out of 1/4 in. Plexiglas and supplied with a clear coolant (50% propylene glycol/H_2O) from a refrigerated circulator. The top need not be sealed if the return outflow is located on a side near the top about 2 ft above the circulator. The outflow fitting should be two times the diameter of the inlet, which should be located on the opposite side near the bottom. Locating a 5/8″ barbed threaded fitting positioned near the top of one side and the inlet a 1/4″ barbed threaded fitting located on the opposite side near the bottom should prevent overflow and give a fairly uniform temperature. If a more uniform temperature is required, the chamber can be positioned on a magnetic stirrer and gently stirred. The pitting of ice can be viewed through the front wall if two small glass Petri dishes are attached to the front and back sides with RTV silicone. Several beads of desiccant should be added before attachment to remove trapped moisture, which will prevent fogging at the low temperature if room humidity is high. A double-walled closed chamber could also be constructed, which would require stirring, and its fabrication is more complicated as well as much more expensive.

2. Put 2 liters of ice flakes in a polyurethane ice "bucket" and add 100 mL of water. Add 12 g of NaCl and stir. Check the temperature after 5 min and it should be around −1.9 °C. If this is less than the freezing point of the culture media, add more salt 0.25 g at a time to lower the temperature. If too cold, add more water. Keep in mind that the viewing tube containing the spent culture media takes only a few minutes to equilibrate before moving it to the refrigerated viewing chamber.
3. Ice crystals can be grown in an insulated "ice bucket." Fill an ice bucket about three quarters full with deionized water and cover. Put in a −20 °C freezer for 12–24 h and remove when there is one half to an inch of ice on the surface. Open a hole on the surface with a hammer, large enough to access the walls, and pour the water out. There should be single crystals protruding from a thin layer of ice on the walls and on the bottom. If no ice has formed on the wall, then it needs to freeze for a longer period of time. These crystals should have a 120° outside angle at the tip with more or less parallel sides. Harvest the ice crystals by cutting their bases with a warm scalpel blade and store in a small jar with a cover that seals. They can be picked up using a small clawed forceps.
4. To determine if a fish is sufficiently anesthetized, it should not respond when the tail is touched. If the fish fails to respond to a tail touch and if it is to be revived, it should be briefly immersed in fresh seawater to flush the MS-222 out of the gills; otherwise, the specimen might not recover.
5. If the initial attempt to pierce the vein is unsuccessful even after probing laterally, then try two or three rays anterior to the failed attempt. The caudal vein becomes larger before entering the kidney near the anal pore and sometime slightly posterior to it. It is important to know where this transition occurs on the ventral side of the fish because puncturing the kidney will not yield any blood.
6. If the serum is to be analyzed for antifreeze hysteresis or for ion levels, it can be stored at +4 °C for a few days. If it is to be kept longer, it is best to store at −80 °C or if collected in the field on dry ice if possible. Long-term storage at −20 °C for several days may cause separation of some of the serum proteins resulting in a large fraction of insoluble protein. It is best to avoid freezing and thawing even if stored at −80 °C because it can also result in solubility problems. This is more prevalent in polar fishes than north temperate fishes probably because of their lipid-rich diet.

7. Depending upon the size of the insect, one might obtain 1 μL or even several microliters. For the common mealworm (*Tenebrio molitor*), often 5–8 μL can be easily collected from either the larva or adult. If an insect is dehydrated as is often the case when collected during the winter, gentle pressure on the thorax may help expel sufficient hemolymph at the puncture site for a hysteresis determination.
8. 5 μL Drummond capillaries can be drawn to a long fine tip using an electrophysiology vertical pipette puller such as a Kopf model 700 C. It can be made fine enough so that it can be inserted into a 10 μL Drummond capillary. Some trial and error is required to adjust the pipette puller heat and pulling force to obtain a long thin tip that will fit inside the 10 μL capillary. The pipette can be inserted into the capillary and 1 μL of sample retrieved. If the sample capillary is not being resealed, the fluid end can also be broken off and the capillary inserted in that end and the movement of the hemolymph meniscus observed, while the pipette is withdrawing the fluid.
9. The freezing point of the culture media can be determined by analyzing a 10 μL sample with a Wescor vapor pressure osmometer or a nanoliter osmometer. The mmol/kg H_2O can be converted to freezing point by multiplying by 0.001858. One mmole/kg H_2O equals 0.001858 °C.
10. Although the growth of ice crystals can be done in either a −20 °C chest or upright freezer, their harvest is best done in a walk-in freezer of −4 to −6 °C, or if not available, it can be done in a −20 °C chest freezer with the cover open because even with it being open, the cold air will remain long enough to shape and attach an ice crystal to the crystal holder. With an upright freezer, the cold dense air quickly flows out of the compartments resulting in melting of the small ice crystals.
11. If a vacuum pump is not available, a vacuum aspirator attached to a water faucet with sufficient flow will generate a similar vacuum. Also 50 mL hypodermic syringes have been used in some cases to pull the necessary vacuum.
12. For some plant leaves, a single round of vacuum infiltration is insufficient and must be repeated several times. Also, to recover the apoplastic wash, the centrifugal force has to be increased. An example is the rice plant where the leaves need longer periods of evacuation, infusion, as well as centrifugation because their stomatal openings are very small and the leaf surface is hydrophobic.

References

1. Scholander PF, VanDam L, Kanwisher JW, Hammel HT, Gordon MS (1957) Supercooling and osmoregulation in Arctic fish. J Cell Comp Physiol 49:5–24
2. Gordon MS, Amdur BE, Scholander PF (1962) Freezing resistance in some northern fishes. Biol Bull 122:52–62
3. DeVries AL, Wohlschlag DE (1969) Freezing resistance in some Antarctic fishes. Science 163:1073–1075
4. DeVries AL, Komatsu SK, Feeney RE (1970) Chemical and physical properties of freezing point depressing glycoproteins from Antarctic fishes. J Biol Chem 245:2901–2905
5. DeVries AL (1971) Glycoproteins as biological antifreeze agents in Antarctic fishes. Science 172:1152–1155
6. Raymond JA, DeVries AL (1977) Adsorption inhibition as a mechanism of freezing resistance in polar fishes. Proc Natl Acad Sci USA 86: 881–885
7. Knight CA, DeVries AL (1989) Melting inhibition and superheating of ice by an antifreeze glycopeptide. Science 245:505–507
8. Cziko PA, DeVries AL, Evans CV et al (2014) Antifreeze protein induced superheating of ice inside Antarctic notothenioid fish inhibits melting during summer warming. Proc Natl Acad Sci USA 110:14583–14588
9. Duman JG (2015) Animal ice-binding (antifreeze) proteins and glycolipids: an overview with emphasis on physiological function. J Expt Biol 218:1846–1855
10. Zachariassen KE, Husby JA (1982) Antifreeze effect of thermal hysteresis agents protects highly supercooled insects. Nature 298:865–867
11. Zachariassen KE, Hammel HT (1976) Nucleating agents in haemolymph of insects tolerant to freezing. Nature 262:285–287
12. Mazur P (1970) Cryobiology: the freezing of biological systems. Science 168:939–949
13. Gruneberg AK, Graham LA, Eves R et al (2021) Ice recrystallization inhibition activity varies with ice-binding protein type and does not correlate with thermal hysteresis. Cryobiology 99:28–39
14. Raymond JA, Sullivan CW, DeVries AL (1994) Release of an ice-active substance by Antarctic Sea ice diatoms. Polar Biol 14:71–75
15. Guo S, Stevens CA, Vance TDR et al (2012) Structure of a 1.5-mDa adhesin that binds its Antarctic bacterium to diatoms and ice. Sci Adv 3:1701440
16. Raymond JA (2011) Algal ice-binding proteins change the structure of sea ice. Proc Natl Acad Sci USA 108(24):E198
17. Bayer-Giraldi M, Weikusat I, Besir H et al (2011) Characterization of an antifreeze protein from the polar diatom *Fragilariopsis cylindrus* and its relevance in sea ice. Cryobiology 62:210–219
18. Raymond JA, Kim HJ (2012) Possible role of horizontal gene transfer in the colonization of sea ice by algae. PLoS One 6(5):E35968
19. Tyler DRV, Bayer-Giraldi M, Davies PL et al (2019) Ice-binding proteins and the 'domain of unknown function' 3494. FEBS J 286:855–873
20. Raymond JA, Janech MG, Fritsen CH (2009) Novel ice-binding proteins from psychrophilic Antarctic (Chlamydomonadaceae, Chlorophyceae). J Phycol 45:130–136
21. Griffith G, Yaish MWF (2004) Antifreeze proteins in overwintering plants: a tale of two activities. Trends Plant Sci 9:399–405
22. Yu XM, Griffith M, Wiseman SB (2001) Ethylene induces antifreeze activity in winter rye leaves. Plant Physiol 126:1232–1240
23. Gentzel I, Giese L, Zhao W, Alonso AP, Mackey D (2019) A simple method for measuring apoplast hydration and collecting apoplast contents. Plant Physiol 179:1265–1272

## Chapter 2

Ice Shell Purification of Ice-Active Compounds

Jessica Morris, Michelle Liddy, and Craig J. Marshall

Abstract

Ice fractionation allows for the selective purification of ice-active compounds from aqueous solutions. A variety of approaches have been developed to allow the interaction of ice with solutions of interest. We describe here a version where a thin layer of ice is formed on the inside of a suitable flask, which is then exposed to solutions containing ice-binding molecules under conditions where the ice shell is allowed to grow slowly. Under these conditions, ice-binding compounds are incorporated into the newly formed ice, and other solutes are accumulated in the remaining liquid fraction.

Key words Ice-active proteins, Antifreeze, Ice nucleation, Ice affinity, Ice shell

1 Introduction

Ice-affinity purification of proteins takes advantage of the property of a class of disparate molecules, mostly proteins, to interact with ice. This is a form of affinity purification but somewhat more general than most such approaches where specific interactions between target and ligand are usual. Unlike most binding surfaces, ice contains a number of potential sites of interaction [1–3] leading to a fairly broad specificity. Ice-binding compounds are quite diverse in their composition and structure, and ice-binding proteins, the most widely studied of these compounds, show a great deal of diversity in size, composition, and structure [4–6]. Ice-affinity purification can be considered as a functional process and little need be known about potential binding compounds or the way they interact with ice. This is helpful when working with uncharacterized sources where little is known about any potential ice-binding compounds.

Supplementary Information The online version contains supplementary material available at https://doi.org/10.1007/978-1-0716-3503-2_2.

Ran Drori and Corey A. Stevens (eds.), *Ice Binding Proteins: Methods and Protocols*, Methods in Molecular Biology, vol. 2730, https://doi.org/10.1007/978-1-0716-3503-2_2,

A number of approaches have been used for ice-affinity purification. The earliest of these is freezing and centrifugation to purify ice-binding proteins from ice algae and plant samples [7, 8]. The technique described here is a modification of that of Kuiper et al. [9] where a cooled finger was placed in a solution containing ice-binding proteins that were incorporated into the growing ice hemisphere. This approach essentially inverts this process where an ice shell formed on the inside of a round bottomed flask is allowed to grow in the presence of ice-binding compounds [10]. Although essentially the same technique, the ice shell is faster and is more scalable than the ice finger and may be more effective due to mixing at the ice-water interface.

The ice shell technique requires little equipment and can be readily scaled to at least 200 mL of extraction fluid at a time. More complex approaches that handle larger volumes are available but typically require more complex equipment [11].

Ice freezing approaches have two advantages in purifying compounds over other techniques. In addition to selectively incorporating ice-binding compounds, the formation of ice largely excludes other dissolved molecules. As a consequence, the ice shell approach can be used for purifying ice-binding compounds from extracts of natural products (such as insect, algal, or plant homogenates) as well as homogenates from bacterial expression systems.

Ice shell growth is influenced primarily by solutes present in the extraction solution and by temperature and can also be affected by antifreeze compounds in the solution. Figure 1b–d shows the effect of salt and flask volume on shell growth, and this may help in choosing suitable conditions. You will need to experiment with your particular conditions.

In addition to osmotically active compounds, solutions for extraction may contain detergents or other additives associated with sample preparation. Our experience is that common detergents such as Nonidet, Triton, and Tween compounds do not cause problems with ice shell fractionation but be aware that removal of detergents from the frozen fraction might alter solubility, and the concentration of detergent in the liquid fraction might cause frothing or unexpected outcomes.

Ice shell growth produces two components: the liquid fraction that is enriched for dissolved compounds including salts, sugars, detergents, and most proteins and other macromolecules and the frozen fraction that comprises mostly water along with a proportion of any ice-binding compounds present in the original solution and which may also contain contaminants that were carried along with ice formation. Any compound that binds ice will be enriched by this process.

Freeze-drying is an effective way of concentrating samples after ice shell fractionation as little salt is likely to be present in the ice shell, particularly after several rounds of purification.

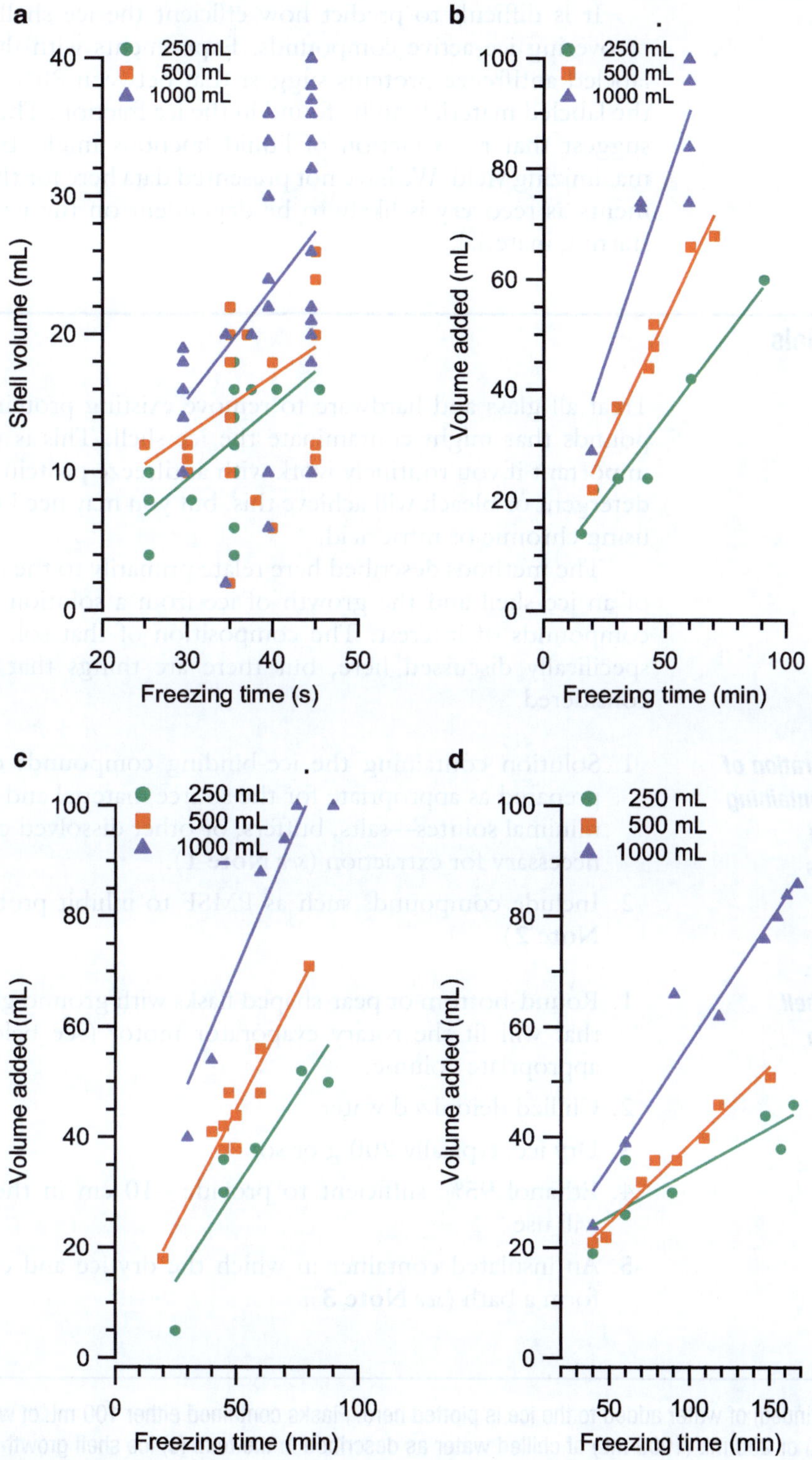

Fig. 1 Ice shell formation and growth for different volumes and ionic strength solutions. (**a**) Ice shell formation from the specified volumes of deionized water. Shell formation was done in an ethanol–dry ice bath as outlined in the text for times between 25 and 45 s. (**b**) Ice shell growth from deionized water in 250 mL, 500 mL, and 1000 mL flasks at −1.5 °C. Shells were allowed to grow for the times specified, and the amount

It is difficult to predict how efficient the ice shell will be in recovering ice-active compounds. Experiments with fluorescently labeled antifreeze proteins suggest that between 30% and 70% of the labeled material can be found in the ice fraction. These findings suggest that re-extraction of liquid fractions might be useful in maximizing yield. We have not presented data here for these experiments as recovery is likely to be dependent on the nature of the starting material.

2 Materials

Treat all glass and hardware to remove existing proteins or compounds that might contaminate the ice shell. This is particularly important if you routinely work with antifreeze proteins. Typically detergent or bleach will achieve this, but you may need to consider using chromic or nitric acid.

The methods described here relate primarily to the preparation of an ice shell and the growth of ice from a solution containing compounds of interest. The composition of that solution is not specifically discussed here, but there are things that should be considered.

2.1 Preparation of Solution Containing Ice-Binding Compounds

1. Solution containing the ice-binding compounds of interest, prepared as appropriate for the source material and containing minimal solutes—salts, buffers, or other dissolved compounds necessary for extraction (*see* **Note 1**).
2. Include compounds such as PMSF to inhibit proteolysis (*see* **Note 2**).

2.2 Ice Shell Preparation

1. Round-bottom or pear-shaped flasks with ground glass fittings that will fit the rotary evaporator motor (see below) of the appropriate volume.
2. Chilled deionized water.
3. Dry ice: typically 200 g or so.
4. Ethanol 95%: sufficient to provide ~10 cm in the bath you will use.
5. An insulated container in which the dry ice and ethanol will form a bath (*see* **Note 3**).

Fig. 1 (continued) of water added to the ice is plotted here. Flasks contained either 100 mL of water (250 mL and 500 mL) or 200 mL (1000 mL) of chilled water as described in the text. (**c**) Ice shell growth from 25 mM NaCl where the conditions of growth were as described for deionized water (**b**). (**d**) Ice shell growth from 150 mM NaCl where the temperature was decreased to −2.4 °C and the other conditions were as for b and c above

6. Measuring cylinder.
7. Timer.
8. Paper towels.
9. Stopper to fit flask opening. Plastic are better than glass and aluminum foil can be used instead.

2.3 Ice Shell Growth

1. Water bath able to cool to at least −5 °C (*see* **Note 4**).
2. Coolant: typically a mixture of car antifreeze (based on ethylene glycol) and water, diluted as indicated on the product (*see* **Note 5**).
3. Motor unit from a rotary evaporator (*see* **Note 6**).
4. Syringe appropriate for the volume of extraction fluid to be used.
5. Tubing that can be attached to the syringe above to allow the solution to be introduced to the flask (*see* **Note** 7).
6. Timer.
7. Pre-cooled extraction fluid.
8. Ice shell prepared as below and stored at −20 °C until needed.
9. Pre-cooled buffer for washing the shell (optional).

3 Methods

3.1 Sample Preparation

1. Keep all solutions on ice to minimize post-extraction proteolysis and enzymatic changes.
2. If the solution contains fragments, filter it through Miracloth or gauze.
3. Clarify the solution by centrifugation at 10,000–20,000× *g* at 4 °C for 20 min.
4. If a lipid layer is present on the top of tube after centrifugation, remove it using a spatula or by absorption onto a piece of clean paper towel (*see* **Note 8**).
5. Once clarified, cool the solution to at least 4 °C and leave on ice until use.

3.2 Ice Shell Preparation

1. The first step of this process is to create an even and thin shell of deionized water around the inside of the flask. The exact volume of this is not important, but for the best results, it should not have any "bare" patches.
2. Add a measured volume of deionized water to the round-bottom flask, cap, and cool to 4 °C. For 250 mL and 500 mL flasks, 100 mL of water is appropriate, and use 200 mL for 1000 mL flasks (*see* **Notes 9** and **10**).

3. Add dry ice to the ethanol bath carefully and allow to cool (*see* **Note 11**). Keep adding dry ice to ensure the bath stays at about −78 °C, the sublimation temperature of dry ice.
4. Start timer.
5. To create the initial thin ice shell, place the flask and water into the freezing bath and tilt to one side. Continually rotate the flask along the axis of the opening at about 60 rpm for between 30 and 45 s (*see* **Note 12** and **Movie 1**).
6. Remove the flask from the freezing bath, and wrap a folded paper towel around the neck to ensure ethanol doesn't drip into the flask or measuring cylinder.
7. Pour the remaining water from the flask into a measuring cylinder and set to one side.
8. Return the flask to the freezing bath and continue to rotate the shell; you will hear loud cracking noises and see cracks in the ice forming inside the flask (*see* **Note 13**).
9. Remove the flask and shell, and either insert a stopper or cap, or cover with aluminum foil, and store at −20 °C until required.
10. Measure the volume removed from the flask and record the difference for later use.
11. Several shells can be prepared in sequence and stored until required. Ensure that they are labeled as it is useful to know the volume of the shell inside each.

3.3 Ice Shell Growth

1. Turn on the cooling bath and allow it to equilibrate to the chosen temperature (*see* **Note 14**).
2. Put the extraction solution on ice and make sure the container opening is easy to access. You may also wish to cool some washing solution at the same time. This can be water or an appropriate buffer solution.
3. Set aside a small portion of the extraction fluid as a control for whatever analysis of the process may be appropriate.
4. Set up the prepared ice shell in the cooling bath angled so that approximately half of it is submerged in the cooling fluid (*see* **Note 15**).
5. Start the rotation promptly so that the ice shell can equilibrate to the bath temperature. You may hear some cracking noises at this stage. The rotation speed should be 30–60 rpm.
6. Remove the tubing from the syringe and gently draw up the cooled extraction fluid until the syringe is nearly full, avoiding air bubbles as these can lead to formation of foam.
7. Attach the tubing to the syringe and insert the tubing down the tube of the evaporator until the end is just above the neck of the flask containing the ice shell.

8. While the flask is rotating, slowly introduce the extraction fluid from the syringe, allowing it to trickle down the side of the flask where it will rapidly equilibrate to the temperature of the cooling bath.
9. Ensure that the ice shell does not start to melt. If it does, you may need to slow the rate at which you add the fluid, or reduce the temperature of the bath (*see* **Note 16**).
10. Once all the fluid is added to the flask, set the timer and allow the ice shell to grow (*see* **Note 17**).
11. To complete the ice shell fractionation, stop the motor and remove the flask from the motor tube. Wipe the flask with paper towel taking care not to warm it and start melting.
12. Record the time.
13. Pour the remaining extraction fluid into a measuring cylinder being careful not to allow any contamination by cooling fluid dripping down the outside of the flask. In a cold room, the flask can be left inverted to drain into the measuring cylinder for about 5 min without the shell melting.
14. At this point, you can assess whether the shell has grown sufficiently (*see* **Note 18**).
15. If the shell has grown sufficiently, you can choose to wash it briefly with 10 mL or so of the buffer set to cool in **step 1** (*see* **Note 19**).
16. Once the ice shell has melted, measure the volume and allow for the water included in the formation of the shell. You can analyze this fraction as appropriate at this point.
17. The liquid fraction can be re-extracted if desired by dilution to the original volume and repeating the ice shell growth. Similarly, the frozen fraction can be re-extracted either directly or after pooling with other ice fractions (*see* **Note 20**).

4 Notes

1. Dissolved solutes affect the freezing and melting point of aqueous solutions, and so the addition of salts and compounds such as glycerol, sucrose, or anything else that affects the colligative properties of the solution should be considered carefully. If the melting point of a solution is reduced by the addition of such compounds, the temperature of the water bath should be lowered, and the time of shell growth may need to be extended. Try to minimize the addition of compounds that will affect the freezing point. Figure 1 gives some information about the effect of NaCl on the temperature and time taken for ice shell growth. In addition, solutions of high protein concentration

may also affect ice formation and may lead to non-specific incorporation of components of the extraction solution. Dilution may be effective at reducing these effects providing that the components of interest are soluble at lower concentrations.

2. If using insect-derived solutions, consider the inclusion of 0.1 mM phenylthiourea from a 100 mM stock dissolved in ethanol to reduce the formation of insoluble melanin compounds. Other inhibitors might be appropriate for your particular application.

3. The container should be in a format that will allow the flask to be rotated with the neck of the flask at 45° to 60° from vertical. As this is a brief process, an insulated container is not strictly necessary, and it can be useful to place the bath on a piece of foam or polystyrene.

4. This process is better done in a cold room but can also be done in the laboratory at room temperature. A cold room reduces the likelihood of ice shells melting while they are being manipulated, and it also reduces condensation from the atmosphere that can form on cooled surfaces and dilute the freezing bath fluid.

5. Ethanol or isopropanol and water can also be used but tends to evaporate with time.

6. Rotary evaporator motor units are typically available freely from Chemistry Departments and the like. For growing an ice shell, all that is needed is the motor to rotate the flask, a stand to allow it to be positioned in the cooling bath, and the glass tubes that give access to the flask through the motor body to allow the introduction of the sample to be fractionated. When choosing a unit, make sure the fittings on the glass tube are compatible with the flasks you have chosen.

7. The tubing should be long enough to extend most of the way through the glass tube and close to the neck of the flask. Ensure that this tubing is thoroughly cleaned between uses.

8. Lipid contamination can also be a problem as the lipids tend to smear on almost all surfaces if not removed. Typically, lipids float on the top of the solution after centrifugation, and these should be removed as much as is practicable.

9. Figure 1a shows how flask size and freezing time influence the volume of the ice shell. Note that there is considerable variation in the volumes especially at longer times and that the volume of the shell does not vary much with the size of the flask.

10. Flask volume is another variable you may wish to change. A larger flask has a larger inside surface area, and so the volume frozen will be greater in a given time: a 250 mL flask has a

surface area of approximately 200 cm^3; a 500 mL flask, 300 cm^3; and a 1000 mL flask, 480 cm^3. We have found that typically an extraction volume of about 20% of the flask volume seems to work fairly well.

11. Be careful when first adding the dry ice as the solution will bubble quite strongly and ethanol may overflow. Be cautious about handling the dry ice as it can "burn" exposed skin. It is also best to do this where there is adequate ventilation as CO_2 gas is released.
12. In our experience, it is important to ensure that shell covers the entire region exposed to water inside the flask and that the thickness of the shell is of less importance.
13. Do not be alarmed by the cracks and pings that you will hear as the shell is fully frozen or by the cracks that appear inside the flask as these are caused by the expansion of water during freezing. Although somewhat alarming when they occur, we have never had an intact flask break during the formation of the ice shell, although flasks with a crack or flaw may break when first introduced to the freezing bath.
14. The growth of the ice shell is affected by several variables. Among these are the volume of the extraction solution, the presence of osmotically active compounds (*see* **Note** 7), the temperature of the cooling bath, and the efficacy of any anti-freeze compounds in the extraction solution. The concentration of dissolved compounds increases as the ice shell grows as most compounds are excluded by the newly formed ice, and this may limit the amount of ice added by lowering the freezing point of the remaining extraction solution. Increased solute concentration may require lower temperatures for ice shell growth. If the cooling bath temperature (and the ice shell) is above the freezing point of the solution, the ice will melt. This can be fixed by reducing the cooling bath temperature or diluting the extraction solution.
15. It is helpful to include a clip to hold the flask to the motor tube as ground glass joints can be temperature sensitive and the flask may detach as it cools and promptly either fill with cooling solution or spill the extraction solution into the cooling bath. Neither is desirable.
16. If melting is significant, you may need to start again.
17. The length of time you grow the shell is a matter of experiment. In our experience, reasonable extraction of ice-binding compounds occurs when about half the extraction fluid is added to the shell. Figure 1b–d provides a guide to how long you may wish to continue the extraction.

18. We have had good results with about half of the extraction solution freezing. If you wish to continue ice shell growth, go back to the start of this section and allow the shell to grow further.
19. You will need to be careful that the ice shell has not warmed sufficiently for it to begin to melt during the wash. Our experience is that draining the ice shell and repeating the fractionation is more effective than washing at removing carried-over material.
20. We have found that two or three rounds of fractionation are usually effective at producing a substantially purified ice-active compound. As noted above, two processes are happening: removal of solutes that are not compatible with ice formation and potential purification of ice-binding compounds. After two or three rounds, solute concentrations will be much lower than at the start, and this will affect the growth of subsequent ice shells and may also affect solubility of compounds in the ice shell fraction.

References

1. Knight CA, DeVries AL, Oolman LD (1984) Fish antifreeze protein and the freezing and recrystallization of ice. Nature 308:295–296
2. Knight CA (2000) Structural biology. Adding to the antifreeze agenda. Nature 406:249–251
3. Knight CA, DeVries AL (1994) Effects of a polymeric, nonequilibrium "antifreeze" upon ice growth from water. J Cryst Growth 143: 301–310
4. Davies PL (2022) Reflections on antifreeze proteins and their evolution. Biochem Cell Biol 100:282–291
5. Davies PL, Graham LA (2018) Protein evolution revisited. Syst Biol Reprod Med 64:403–416
6. Bar-Dolev M, Braslavsky I, Davies PL (2016) Ice-binding proteins and their function. Annu Rev Biochem 85:515–542
7. Raymond JA, Fritsen CH (2001) Semipurification and ice recrystallization inhibition activity of ice-active substances associated with Antarctic photosynthetic organisms. Cryobiology 43: 63–70
8. Raymond JA (2000) Distribution and partial characterization of ice-active molecules associated with sea-ice diatoms. Polar Biol 23: 721–729
9. Kuiper MJ, Lankin C, Gauthier SY, Walker VK, Davies PL (2003) Purification of antifreeze proteins by adsorption to ice. Biochem Biophys Res Commun 300:645–648
10. Marshall CJ, Basu K, Davies PL (2016) Ice-shell purification of ice-binding proteins. Cryobiology 72:258–263
11. Adar C, Sirotinskaya V, Bar Dolev M, Friehmann T, Braslavsky I (2018) Falling water ice affinity purification of ice-binding proteins. Sci Rep 8:11046

Chapter 3

Ice Isn't the Only Crystal in Town: Structure Determination of Ice-Binding Proteins via X-Ray Crystallography

Tyler D. R. Vance

Abstract

Ice-binding proteins (IBPs) are proteins that have the remarkable ability to bind to ice, and their study has intrigued researchers for decades. This chapter explores the importance of structural biology in understanding IBPs and highlights the significant contributions of IBPs to the field of structural biology. The structures of various IBPs from different organisms have been elucidated, revealing key elements involved in ice binding. Structural biology techniques, including nuclear magnetic resonance (NMR) spectroscopy, transmission electron cryo-microscopy (cryo-EM), and X-ray crystallography, play crucial roles in solving protein structures. This article focuses on X-ray crystallography as a tool for investigating IBP structures, providing insights into its theoretical and practical aspects, experimental workflows, and common pitfalls to avoid.

Key words Ice-binding proteins, IBPs, Structural biology, Protein structures, NMR spectroscopy, Cryo-EM, X-ray crystallography, Ice binding, Protein folding, Protein tertiary structure, Experimental techniques

1 Introduction

There are proteins that bind to ice. To researchers in the field of ice-binding proteins (IBPs) and other readers in the know, the preceding sentiment is hardly a surprise; indeed, the existence of ice-binding—originally denoted as antifreeze—proteins was first published in the early 1970s, and their study has continued onward ever since. But to the majority of humanity that does not know about ice-binding proteins—which includes both laymen and avid protein biochemists alike—the personal discovery that such proteins exist is a thing of awe. Awe, followed swiftly by the age-old question: how? How are proteins—these mixtures of carbon, nitrogen, oxygen, and a little sulfur for good measure—able to adhere to frozen water?

Ran Drori and Corey A. Stevens (eds.), *Ice Binding Proteins: Methods and Protocols*, Methods in Molecular Biology, vol. 2730,
https://doi.org/10.1007/978-1-0716-3503-2_3,

Discovering the answer required many years of experimental work from a variety of different disciplines, including the keystone discoveries brought forward by structural biology: the elucidation of macromolecular structure at atomic-level resolution. To this day, structural biology continues to play a role in our growing understanding of IBPs, making it a valuable tool for any lab interested in these awe-inspiring macromolecules and how they work.

1.1 The Importance of Structural Biology to Ice-Binding Protein Research

Structural biology of ice-binding proteins began in 1995, when the type I antifreeze protein (AFP) from winter flounder (*Pseudopleuronectes americanus*) became the first ice-binding protein structure to be solved [1]. Since then, researchers have elucidated the structures of 12 different types of IBP folds hailing from fishes, insects, plants, fungi, and bacteria (*see* Fig. 1a) [1–14]. The diversity of their source organisms is mirrored by their divergent tertiary folds (*see* Fig. 1b).

Through this myriad of different structures, our current understanding of how IBPs bind to ice has solidified. The structures revealed the protein elements required for forming an anchored clathrate of water that can, in turn, fuse into a growing ice crystal [9]. These required elements include (1) a flat, hydrophobic surface that serves as the main source of entropic energy for structuring ice-like waters and (2) hydrogen bond-forming residues that line the site to anchor the ice-like waters to the protein. IBP structures have also served as the foundation for molecular dynamic experiments that attempt to simulate the kinetic/thermodynamic parameters of ice binding.

Structural biology has helped the field of IBPs, but IBPs have also helped the field of structural biology. As the majority of IBPs are produced via convergent evolution (save for the DUF3494-containing ice-binding proteins of fungi, algae, and bacteria), the majority are structurally distinct from each other and—in several cases—all other known folds. Indeed, the discovery of many IBP structures has heralded the addition of a new protein tertiary fold to the ever-expanding databank of protein structures. Examples of unique structures derived from the study of IBPs include the insect IBPs with internal disulfides throughout the coil (e.g., *Tenebrio molitor* IBP and spruce budworm IBP) and the hyperactive type I isoform "Maxi" with water molecules coordinated within a hydrophilic core.

As research into IBPs continues, the discovery of new IBP structures has implications within and beyond the field of cryobiology. But how does one go about solving a protein structure?

1.2 The Big Three of Structural Biology

When it comes to solving protein structures, there are three main techniques currently in use: (1) nuclear magnetic resonance (NMR) spectroscopy, (2) transmission electron cryo-microscopy

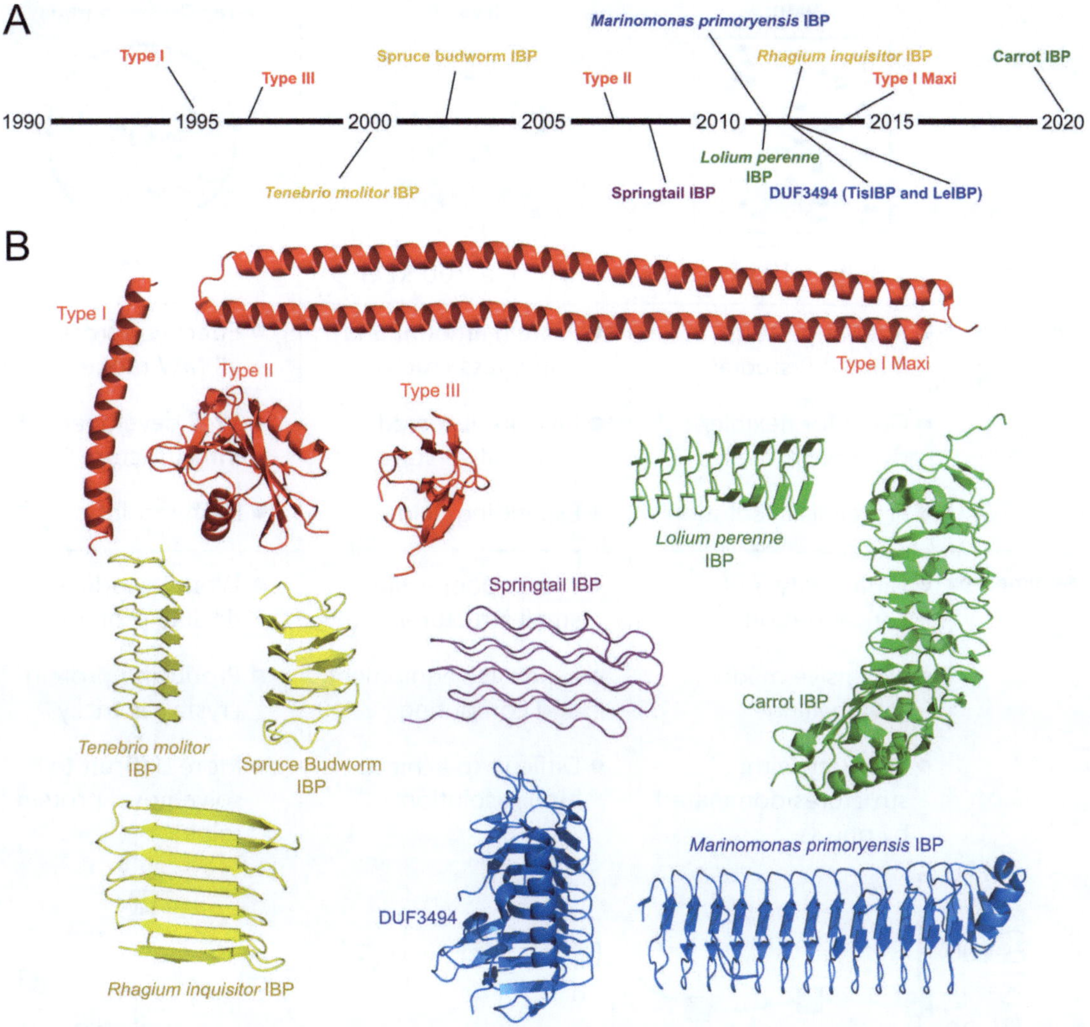

Fig. 1 IBP structures. (**a**) Timeline for the discovery of new IBP folds (fish AFP = red; insect IBP = yellow; non-insect arthropod IBP = purple; plant IBP = green; microorganism IBP = blue). (**b**) Fish IBPs (red) range in structure from the entirely alpha-helical type I AFPs to globular small folds like that of types II and III. The springtail arthropods produce IBPs (purple) dominated by poly-proline helices. Meanwhile, the majority of insect (yellow), plant (green), and fungal and bacterial (blue) IBPs use alterations on the beta-solenoid fold: a spiraling coil of parallel beta strands that take up various shapes and lengths

(cryo-EM), and (3) X-ray crystallography. All three are valuable techniques, each possessing their own strengths and weaknesses (*see* Fig. 2).

First, we have NMR spectroscopy, which subjects proteins in solution to a constant magnetic field and then measures their interaction with radio waves. The wave frequencies that are absorbed by the magnetized sample are determined by the specific chemical environments (i.e., spatial proximity and chemical connectivity) of the atomic nuclei within the protein. Therefore,

	NMR	Cryo-EM	X-ray Crystallography
Technique	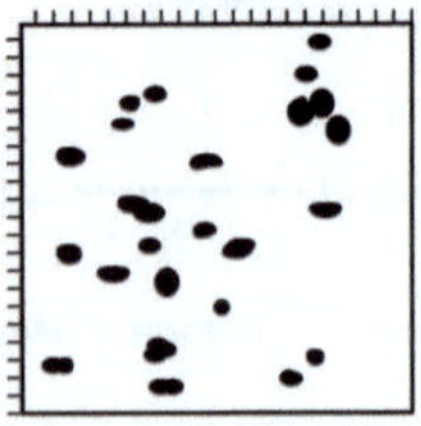	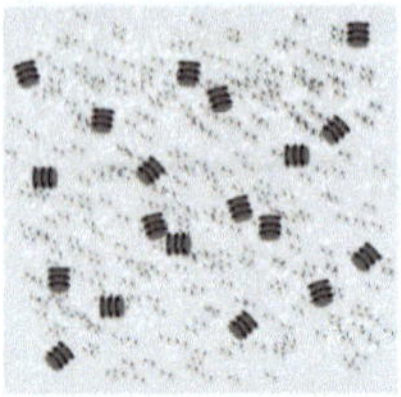	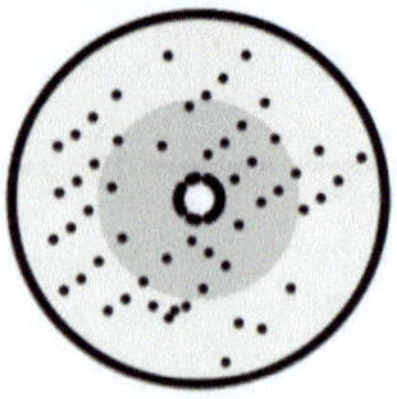
MW Range*	Below 40 kDa	Above 100 kDa	N/A
Strengths	• Size range suits many IBP studies • Good for flexible/ dynamic proteins • Protein is in solution	• Protein amount and purity less crucial • Protein observed in its native state • Expanding field	• Effective across all MW ranges • Well developed infrastructure • High resolution
Weaknesses	• High purity and concentration • Expensive media for labelling • Difficulty with structures dominated by one ss	• Works poorly with small structures • Expensive equipment and computing • Difficult to achieve high resolution	• Works poorly with flexible proteins • Producing protein crystals is tricky • More difficult to solve novel protein folds

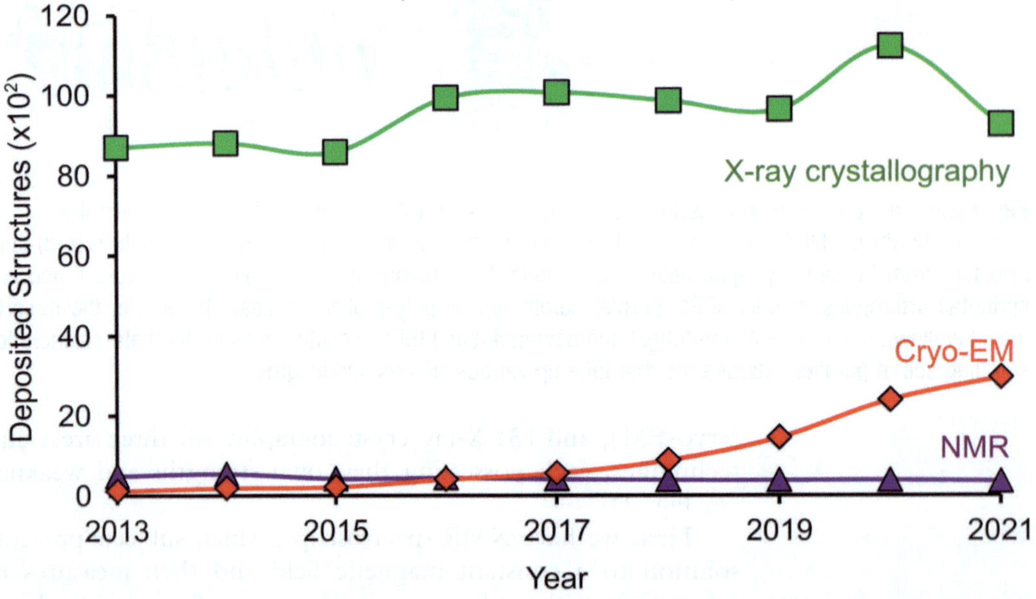

Fig. 2 Comparison of structural biology methods. A table of comparison for NMR, cryo-EM, and X-ray crystallography. *MW ranges are most likely scenarios; examples of solved proteins outside of these ranges are sparse but do exist

protein structural data can be elucidated through the accumulation of "chemical shifts" in excitation frequency for each nucleus within the protein.

As the only structure determination technique that deals with proteins in solution, NMR is uniquely capable of studying dynamic protein systems, such as intrinsically disordered proteins. However, resolving the sheer number of often overlapping nuclei signals can be incredibly difficult if not impossible, a problem that compounds with larger molecules. As such, NMR's use is limited to smaller proteins (below 40 kDa is a necessity, though most solved NMR structures fall far below this cutoff).

In terms of use with IBPs, the size limitation is not a huge detriment as most IBPs are below 40 kDa, with the fish AFPs (several of which were solved via NMR spectroscopy) being perfectly sized at below 10 kDa. However, proteins where many of the nuclei are in similar chemical environments (such as IBP structures dominated by a single secondary-structure element) can result in obfuscating clusters of chemical shifts. More generally, the expensive media used for NMR-labeling proteins during expression and the especially stringent purity requirements have kept NMR as a niche technique, with the lowest number of deposited structures in the Protein Data Bank (PDB) in recent years [15].

Second, cryo-EM is a form of transmission electron microscopy where electrons are flooded through a thin layer of vitreous ice that has macromolecules (such as protein) suspended within. Electrons interact with the sample, producing an image that is captured via a direct electron detector. However, the image is only a two-dimensional snapshot; to achieve a full three-dimensional structure, many images of the protein in multiple orientations must be taken and combined. Depending on quantity and quality of the images captured, the resulting structures can rival the resolution of the other two methodologies.

Due to the imaging and selection of single particles in cryo-EM, it is possible to resolve multiple conformations and filter out contaminating macromolecules within a sample, making cryo-EM the least stringent technique in terms of protein purity, as well as the least demanding in terms of protein quantity. This combined with the ability to capture proteins in their native (albeit frozen) state makes cryo-EM a very enticing technique; indeed, more and more structures solved by cryo-EM are being deposited to the PDB each year [15].

However, EM images are plagued by high background noise, which can make detecting the shape and features of proteins difficult. This problem is aggravated when molecular weight decreases, as smaller proteins produce a poorer signal-to-noise ratio, making IBP structural studies with cryo-EM unlikely. Even though select best-case scenarios have achieved atomic resolution of ~40 kDa proteins, most structural studies using cryo-EM are currently limited to larger proteins and complexes above 100 kDa; with the exception of the ice-binding adhesins, no known IBP reaches this size cutoff.

Last but certainly not least, X-ray crystallography is the original method of protein structure determination and remains the most readily used. The technique has no explicit size limitations, is able to achieve the highest resolutions possible, and—as the elder statesman of structural biology—boasts an impressive repertoire of tools and infrastructure to ease the collection/processing of data. Given these boons, it is not surprising that the majority of IBP structures have been solved using this method. However, just because it is the most readily available tool in the box does not make it any less difficult to master.

As such, the rest of this chapter will focus on X-ray crystallography as a tool for solving the structures of IBPs. The goal is to introduce readers to the basics (both theoretical and practical) of the technique, establish a workflow for a crystallography experiment, and offer advice on common mistakes to avoid.

2 Theory of X-Ray Crystallography

If you were to walk into a structural biology lab in the modern era, you very well might see a scientist hunched over a microscope with their face pressed hard against the eyepiece, breath held in anticipation, and hands steadily holding a pencil-like appendage with its tip periodically dipping in and out of a many-welled plate. What you are watching is an attempted extraction of a protein crystal from the solution that birthed it. As you observe, you will (hopefully) see the held appendage be suddenly lifted from the plate and—after a small detour into a droplet of liquid mere microliters in volume—submerged into a vat of liquid nitrogen. From there, the frozen crystal will be sent to a facility known as a "synchrotron," where a combination of on-site scientists and remotely controlled robot arms will assist the intrepid structural biologist in collecting a series of images containing shifting patterns of black dots. Then, if you have the time and patience to maintain your vigil, you might be lucky enough to see all of those images compiled and distilled into a single file that—after being run through a myriad of strange programs with names like Aimless, Phenix, and Coot—produces a rough semblance of what looks to be a protein structure.

This whimsically vague explanation of a generic X-ray crystallography workflow is but the surface of the impressively deep well of scientific theory and practical prowess that lies beneath every solved protein structure. Thanks to X-ray crystallographers from generations past and the many automated tools they produced, the majority of crystallographic experiments can be done without an extensive spelunking into the theory. That said, a foundation is needed to understand why the actions listed above are necessary for solving a protein structure. This section aims to provide a cursory explanation of the theory behind X-ray crystallography. For further reading, look to such lauded texts as "X-ray structure determination: A practical guide" by George H. Stout and Lyle H. Jensen and "X-ray crystallography" by Gregory S. Girolami.

2.1 X-Ray Diffraction: The Closest We Have to an X-Ray Camera

Why can't we see a protein? While the answer seems fairly obvious (i.e., it is too small), the question is a useful one when discerning why we use X-rays for determining protein structures. Our eyes take in light within the aptly named "visible" portion of the electromagnetic spectrum that has reflected off of objects in our field of view. Visible light has a wavelength between 400 and 700 nm and can therefore interact with matter close to or larger than that size. Singular proteins are much smaller than the wavelength of visible light, and their features—as defined by atoms connected via covalent bonds—are even smaller. Luckily, X-rays have a wavelength of between 0.01 and 10 nm, the perfect size for interaction with covalently bonded atoms (within the realm of 0.1 nm), so all we need is a lens that can refract X-rays into a focused image like our eyes do for visible light and voila: a protein. If only such a lens existed, for X-rays are unable to be focused due to their refractive index being very similar across all known materials.

In lieu of an X-ray lens to solve all our problems, X-ray crystallography relies on the scattering of X-rays upon interaction with electrons. When an incident beam of X-rays hits an atom, electrons scatter the radiation outward in all directions. Given how many electrons there are in a single protein, a complex mixture of scattered X-ray waves is produced, with the waves interacting with each other both constructively (increasing amplitude) and destructively (decreasing amplitude). The resulting pattern of high-intensity and low-intensity regions is indicative of the electron positioning (i.e., the protein structure) that produced it and is known as X-ray diffraction. The production of constructive interference patterns is shown in a simple two-electron system in Fig. 3.

From the above description, it seems like you should be able to take a solution of protein, place it in front of an X-ray beam, and get yourself a structure. Sadly, this is not the case, as proteins in solution are too flexible and mobile to produce a consistent pattern; X-rays shot into a protein solution would scatter against all the different conformations and orientations of your protein. Even if you were somehow able to rustle a single protein into position, you wouldn't find a particularly strong pattern awaiting you on the other side. To solve this problem, it is time for the latter half of "X-ray crystallography" to come into play.

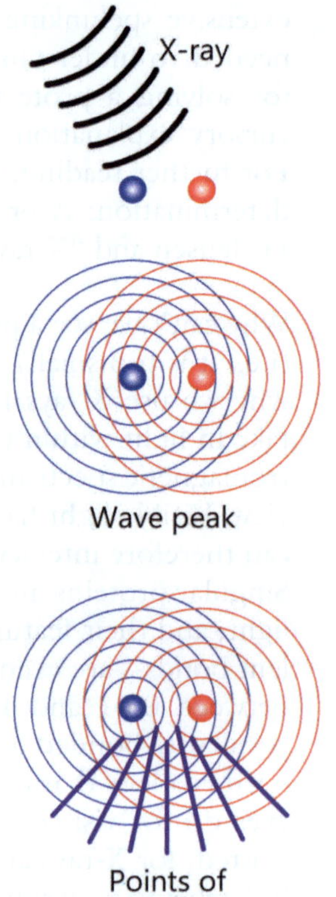

Fig. 3 X-ray diffraction patterns are made by constructive interference. When two electrons (blue and red dots) interact with an incoming X-ray, they each spherically reflect the radiation outward. Where these reflections overlap, a point of constructive interference is produced

2.2 Protein Crystals as Signal Amplifiers

A protein crystal is effectively an incredibly ordered form of aggregation, where the macromolecules precipitate out of solution by preferentially forming contacts with copies of itself rather than solvent. What makes crystals different than plain old aggregation is that every molecule is in the same orientation and making the same contacts, forming a solid mass of protein organized into repeating three-dimensional patterns called lattices. To provide some terminology, a crystal lattice is divided into unit cells, which are the smallest portions of the crystal that can be translated to make every other portion; similarly, unit cells are made up of asymmetric units, which are the smallest portion of the unit cell that can be rotated to make every other portion. In the case of protein crystals, an asymmetric unit contains one or more copies of the protein of interest.

Crystals can be defined through a combination of their unit cell parameters—i.e., the length, width, and height of the unit cell and the angles of its sides (*see* Fig. 4a)—and its space group. A space

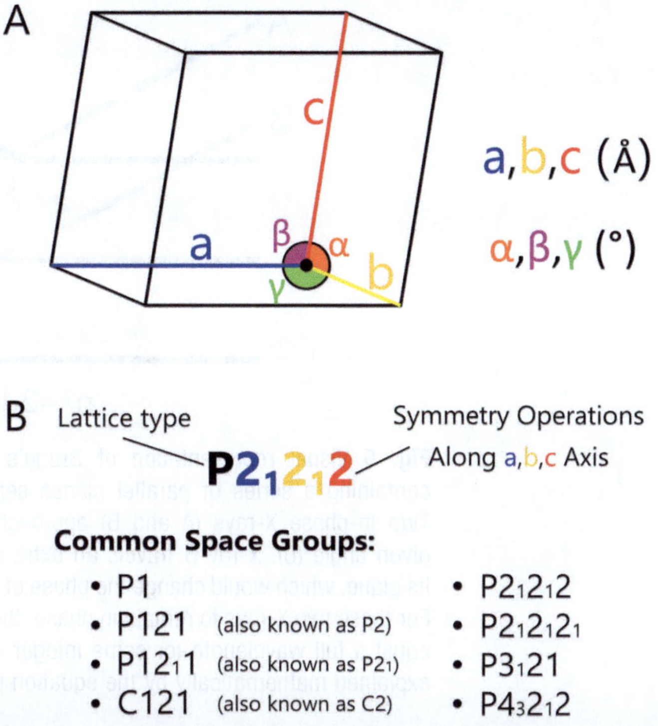

Fig. 4 Parameters that define a crystal. (**a**) Unit cell parameters include the lengths of the three axes (a, b, c) and the magnitude of the angles between them (α, β, γ). (**b**) The space group outlines how the molecules in the crystal are symmetrically related. Space group notation has two components: the lattice symbol (e.g., P, C, I) which indicates the type of face centering and the list of symmetry operations that occur along the a, b, and c axes

group is a roadmap to where proteins are found in the crystal, providing the symmetry operations necessary to reproduce the unit cell (and by extension the entirety of the crystal) from a single asymmetric unit. While many space groups are possible, there are a few that show up in protein crystallography more frequently (*see* Fig. 4b) [16].

When X-rays hit a protein crystal, the resulting diffraction pattern is amplified and sharpened relative to that of a single protein. The points of constructive interference can now be registered by an X-ray detector, where they appear as discrete spots. The intensity of the spot is determined by the amplitude of the constructively produced wave that made it, which is in turn determined by the electron density of the protein. However, where the spots appear on the detector also gives information about how the proteins are placed within the crystal lattice. To figure out why, we need to delve into two mathematical constructs and the interplay between them: Bragg's law and the reciprocal lattice.

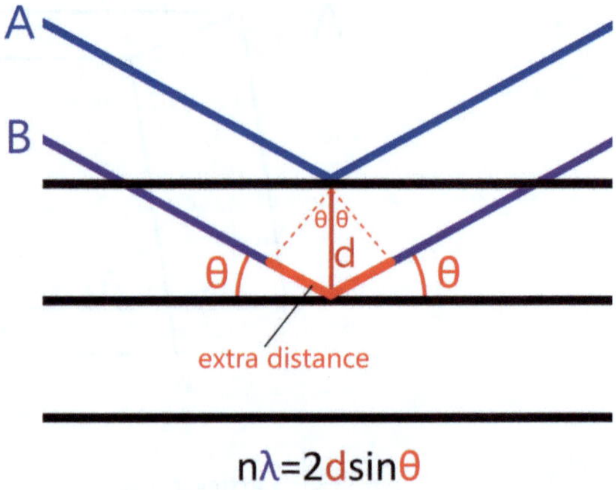

Fig. 5 Visual representation of Bragg's law. The crystal is visualized as containing a series of parallel planes separated by a constant distance (d). Two in-phase X-rays (A and B) approach and reflect off these planes at a given angle (θ). X-ray B travels an extra distance (red) to make contact with its plane, which would change the phase of these two rays relative to each other. For these two X-rays to remain in phase, the extra distance X-ray B travels must equal a full wavelength (or some integer of wavelengths). This relationship is explained mathematically by the equation provided

2.3 Simple Explanations for Non-simple Math

No discussion of X-ray crystallography would be complete without a look at Bragg's law. It is a simple and elegant mathematical formula that explains the spatial relationship necessary for reflected waves to stay in phase. As applied to X-ray crystallography, Bragg's law interprets the crystal as a series of parallel planes separated by a distance (**d**). In Fig. 5, two X-rays (labeled **A** and **B**) are in phase before they contact the planes. If they are to remain so, B must travel an extra distance equal to some integer of full wavelengths. This relationship is summarized in the provided equation (*see* Fig. 5).

While many conceptual parallel planes could be drawn through a crystal, only the ones that satisfy Bragg's law will result in constructive interference. This means that every set of Bragg-approved parallel planes can be linked to a potential spot of X-ray diffraction (hence why the spots are often referred to as "reflections"). The Bragg's law equation shows an inverse relationship between the angle of reflection and the distance between the planes, meaning that smaller distances between planes will result in a larger angle of reflection. Thought of in another way, the closer the electrons are to each other, the further away from the center of the detector their resulting spot will be. Hence, "high-resolution data" that hold information capable of resolving close together atoms are found at the edges of the detector in crystallographic experiments.

In general, the position of spots on the detector is determined by the periodicity of the crystal lattice (namely, unit cell parameters and space group) and can be visualized as an array of imagined points in space through a mathematical transformation of the crystal lattice into a reciprocal lattice. If you could see this imaginary mathematical construct known as the reciprocal lattice, it would appear as an array of dots spreading out from the crystal in all directions. Each of these dots is a potential diffraction spot, with an unknown intensity that is determined by the electron density of the protein. The goal of an X-ray crystallography experiment is to sample as many of these reciprocal lattice points as possible with the X-ray detector. This is done by rotating the crystal, which in turn rotates the reciprocal lattice points to bring more of them in contact with the detector over the course of the experiment.

With the theory well in hand, we can now look to its application throughout the X-ray crystallography workflow, which can be broken into three parts: (1) sample preparation and crystal growth, (2) data collection, and (3) structure solution and refinement.

3 Sample Preparation and Crystal Growth

Proteins are incredibly complex macromolecules that possess a huge range of biochemical properties that, in turn, could produce any number of potential structures. This variety is what allows proteins to perform so many different tasks within biological processes, including the astonishing ability to bind to ice; it is also what makes determining a method for crystallizing a protein so difficult, as no two proteins are truly the same. This section will focus on how to consistently produce crystals for structure determination studies, starting with the important steps of protein expression and purification.

3.1 Protein Samples: Crap In, Crap Out

This section's title is a blunt-but-effective phrase with an obvious meaning: the output from your experiments (i.e., your crystals and resulting structure) is hampered by the quality of your input (i.e., your protein sample). The ideal protein sample for an X-ray crystallography experiment is one that is expressed in large quantities, readily purified to homogeneity, monodispersed in solution, and well-folded with limited flexibility. While clearly not all proteins will be able to reach this high bar, there are certain requirements that can be bent without breaking and others that cannot.

Do I have enough protein? Crystallography experiments can require a large amount of sample. Of course, the actual concentration of protein that yields a crystal differs between proteins; some can crystallize at concentrations below 2 mg/mL, yet others require upward of 50 mg/mL. While each crystallography lab has its own practices, a good starting point for protein concentration is ~10 mg/mL. The volume of 10 mg/mL protein required is dependent on the setup of your crystallization experiments, which will be discussed in Sect. 3.2.

Is my protein pure enough? The purity of a sample is not as important for crystallography as it is for other techniques like NMR. After all, crystallization is effectively a method of purification. That said, contaminating proteins can interfere with the crystallization of your protein of interest by making unwanted contacts with your protein or changing the solute–solvent dynamics of the solution; worse, the contaminants could crystallize and send you down a feverish goose chase that ends with a beautiful 2 Å structure of an already known *E. coli* protein [17]. For these reasons, it is recommended that your protein sample be at least 90% pure, as determined by SDS-PAGE. While IBPs present a unique form of affinity purification that should ease this process (see Ice-Affinity Purification), the 90% level of purity applies to both purification from contaminating proteins and purification of your protein to a single population. If there is variation in post-translational modifications, this can greatly impact reproducibility of crystal formation.

Is my protein "well behaved"? Biochemists often anthropomorphize proteins as children, with the term "well behaved" used to denote proteins that refrain from aggregation (monodispersed) and maintain a consistent (preferably rigid) fold. Both aspects are very important for crystallography. If the protein is already precipitating in its base state, it is very unlikely that it will be convinced to halt that aggregation for the more ordered variation of crystallization. Likewise, if the protein contains highly flexible portions (either intrinsically or from a lack of proper folding), a rigid crystal lattice is highly unlikely.

Determining if a protein is well behaved or not can be done through many different biophysical techniques. Monodispersity can be tested via tried-and-true light scattering techniques such as dynamic light scattering (DLS) or size-exclusion chromatography-linked multi-angle light scattering (SEC-MALS) or through newer systems such as the mass photometer. A size-exclusion chromatography column properly calibrated with standard proteins of known size can be used as a less specialized alternative. Proper folding can be inferred through dyes that bind hydrophobic patches on proteins, tryptophan fluorescence experiments, and circular dichroism spectroscopy. The subsequent stability of these folds can be tested through temperature ramping experiments to see how long the structure is maintained before unfolding or "melting" occurs.

Ways to prevent or mitigate bad behavior in proteins will be discussed in Sect. 3.3.

3.2 The Black Magic of Crystal Growing

As stated above, protein crystallization is a form of highly organized aggregation. Therefore, many of the compounds capable of aggregating proteins and ruining a biochemist's day find a new calling as crystallization conditions—different pHs, salts (e.g., ammonium sulfate), and precipitant polymers (e.g., polyethylene glycol or PEG) among them. The difficulty is in determining which combination of which compounds at what concentration will produce a protein crystal.

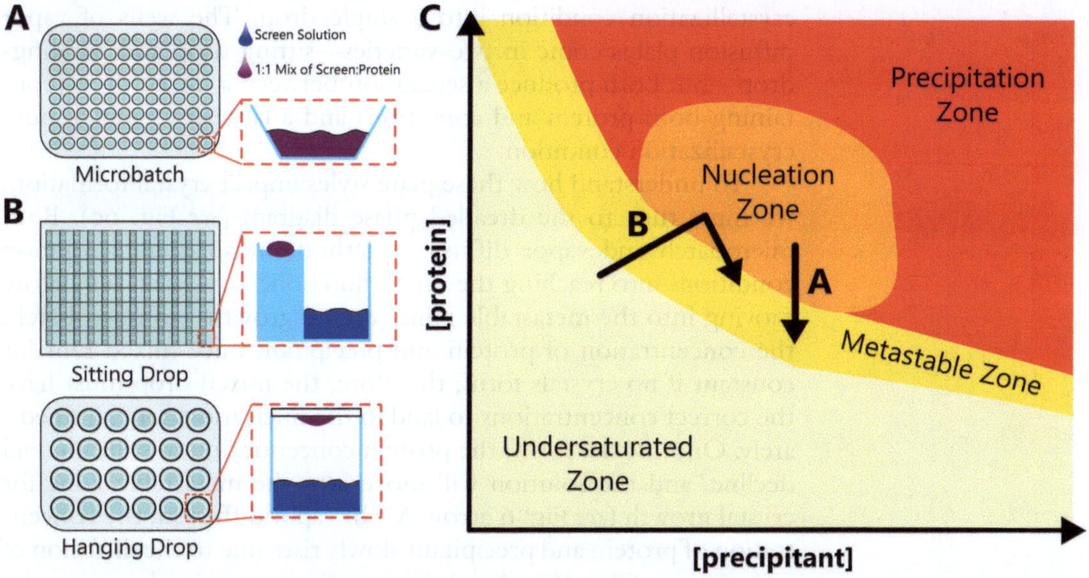

Fig. 6 Plates and phase diagram of protein crystallization. Standard plates used for crystallography experiments include both microbatch (**a**) and vapor diffusion (**b**) examples. Vapor diffusion can be further separated into sitting- vs. hanging-drop trays. (**c**) A protein crystallization phase diagram explores how the relationship between protein concentration and concentration of precipitant impacts crystal formation. The diagram can be separated loosely into an area of undersaturation, where low protein and/or precipitant concentrations keep protein in solution, and an area of supersaturation, where high protein and/or precipitant concentrations result in aggregation of protein. The area of supersaturation can be further broken up based on the manner of aggregation. The precipitation zone yields random and unordered aggregation. The nucleation zone yields an initial ordered clustering of proteins (i.e., a crystal nucleus), which is the biggest energetic hurdle to overcome in crystal formation. The metastable zone provides the right conditions for the growth of a crystal nucleus into a full-fledged crystal. The boundaries of these zones can be changed through altering solution conditions (e.g., pH, salts, small-molecule and detergent additives, temperature)

While many attempts have been made to predict crystallization conditions, the truth is that growing protein crystals is still very much a numbers game. The process involves dispensing your protein of interest into large arrays of multi-component solutions that have previously produced crystals. The more solutions tried, the greater the chance of encountering one that produces a crystal. These conditions have been conveniently collected into commercially available screening kits (e.g., PACT, PEGs, Top96, JCSG+, etc.), which are used by mostly all crystallography labs.

Mixing your protein with these crystallization conditions occurs in multi-well plates. There are many commercially available plates, though they all cluster into one of the two types (*see* Fig. 6): microbatch (**A**) or vapor diffusion (**B**). The wells of microbatch plates are simple plastic basins in which you mix protein and crystallization condition into a single drop. The wells of vapor diffusion plates come in two varieties—sitting-drop and hanging-drop—but both produce a separation between a mixed drop (containing both protein and condition) and a large reservoir of pure crystallization condition.

To understand how these plate styles impact crystal formation, we must turn to the dreaded phase diagram (*see* Fig. 6c). Both microbatch and vapor diffusion methods must arrange solution conditions into reaching the nucleation zone (crystal start), before moving into the metastable zone (crystal growth). In microbatch, the concentration of protein and precipitant once mixed remains constant if no crystals form; therefore, the mixed drop must have the correct concentrations to land in the nucleation zone, immediately. Once nuclei form, the protein concentration in solution will decline, and the solution will move into the metastable zone for crystal growth (*see* Fig. 6 arrow A). In vapor diffusion, the concentration of protein and precipitant slowly rises due to the diffusion of water vapor from the mixed drop to the reservoir; therefore, the initial mixture does not need to be in the nucleation zone. Instead, the drop will increase in concentrations into the nucleation zone before the decline in protein concentration via nucleation once again brings the drop into the metastable zone (*see* Fig. 6 arrow B).

Moving beyond the theoretical crystallography landscape of a phase diagram, Protocol 1 provides an example of a standard crystallization experiment. However, it is unlikely that such screening will produce singular, large, well-diffracting crystals that can just be thrown into your X-ray beam of choice. Often, screening produces "hits," which are indications that crystals can be formed within the condition if given a little optimization. Promising hits include tiny microcrystals that refuse to grow or intense clusters of crystal-like aggregates that form black spiky balls (referred to as sea urchins). Less promising examples include phase separations—distinct drops of concentrated protein within the drop—and highly localized (referred to as "good") precipitate.

Hits can hopefully be optimized to produce usable crystals through one of the several methods or some combination therein: (1) optimizing the condition through an array of slight alterations to precipitant concentration, pH, or protein concentration that is often a good start; (2) additive screens that test a multitude of small molecules and detergent additions to the condition; and (3) seeding experiments where a promising hit is ground into small microcrystalline seeds and then distributed either into alterations of the original condition (classical microseeding) or into an array of screening conditions (random microseeding matrix screening or rMMS) [18].

3.3 If the Worst Should Happen...

In the event that no hits are found after trying several screening suites, there are several next steps to consider.

First, check how many wells in your crystallization screens have obvious precipitate in them. Precipitation, as foretold by the phase diagram, indicates that the nucleation zone has been overshot. If there are too many wells with precipitate (i.e., over 60% of tested conditions), the protein concentration may be too high; alternatively, if there are too few wells with precipitate (i.e., under 40% of tested conditions), the protein concentration may need to be increased.

If the protein concentration produces precipitate within that 40–60% of conditions, the issue may be the protein itself. Try removing (or, alternatively, retaining) solubility and purification tags, subjecting your protein to glycosylases to cleave off unwanted post-translational glycosylations, performing limited proteolysis to selectively clip exposed and flexible portions of the protein, or perhaps co-crystallizing your protein with known binding partners or immunoglobulin-derived molecules (e.g., Fabs, nanobodies).

If these fail, then it is time to go back to the DNA construct and try alternatives. Indeed, most crystallography labs begin studying new proteins by trying several N- and/or C-terminal truncations, along with the full-length protein. Choosing which truncations to make can be difficult if a primary sequence is all that is known about the protein's structure. Fortunately, structure modeling software like Phyre2, Robetta, I-TASSER, and AlphaFold are often able to provide guidance as to where domains begin and end, allowing researchers to exclude flexible tails from their protein constructs. If these programs fail, secondary structure prediction can provide clues as to where unstructured (and likely flexible) portions of the protein reside.

If even these planned truncations fail, perhaps the best advice that can be given is to give up. Try a different structure determination methodology, try a different protein, try a different research project. The sunk cost fallacy has claimed many structural biologists, and there are still many fruits left on reachable branches of the tree of life.

4 Data Collection

We have now worked our way back up to the hypothetical crystallographer in Sect. 2 with their plastic tray full of crystals. However, though growing crystals is often the biggest bottleneck in structure determination via X-ray crystallography, it is far from the end of the process. The next step is the actual collection of diffraction data, where the crystal is removed from the plate (Protocol 2) and bombarded with X-rays so that the constructive interference pattern produced by the interaction between X-rays and electrons can

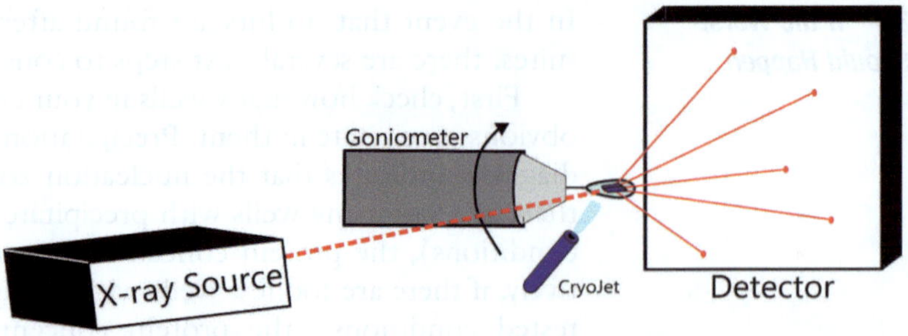

Fig. 7 General setup for an X-ray diffraction experiment. An X-ray source is projected toward a detector, with the crystal placed in between. The crystal is suspended in liquid within a small loop, which is in turn placed on a goniometer that is able to rotate the crystal within the X-ray beam. A cryojet can be used to keep the crystal cool (100 K), thereby reducing radiation damage. With this setup, protein crystals can be easily rotated to sample the most diffraction spots possible

be recorded and deciphered. To collect X-ray data, an X-ray source and a detector are required (*see* Fig. 7). The crystal is placed between the two on a rotating goniometer. A series of "frames" are then collected, where each frame is a single diffraction image taken over the course of a set rotation increment of the crystal. A complete dataset will contain many frames, the culmination of which samples a large portion of the reciprocal lattice of the crystal.

4.1 Different Strokes: The Various Ways of Collecting Data

With this near-universal setup, there are a variety of different types of collection experiments researchers can run.

First, there are screening experiments, where minimal data (only a few frames) are collected on a number of crystals in sequence for the express purpose of gaining information about the crystals. This could be information on the space group and unit cell dimensions, the quality of the diffraction spots, or the maximum resolution that the crystals will be able to reach. Of course, diffraction screening can also answer the age-old question "is it actually protein?" Many crystals that form during crystallography experiments are not protein at all, but salt crystals. There are a variety of methods for testing if a crystal is salt or protein, such as stains like Izit dye that can only enter the large solvent channels of protein crystals or the "crush test" where crushing the crystal with a metal/glass rod indicates if it is protein (soft and gelatinous) or salt (hard and brittle). But perhaps the most definitive test is with X-ray diffraction, as protein will produce a pattern of many dots across the detector and salt will only produce a few high-resolution spots of high intensity.

Following screening, the "best" crystals (see Sect. 4.2 for a more apt descriptor than "best") are chosen for native collection experiments. These experiments are the standard for collection, with the goal being to acquire (1) many frames of data that cover as much of the rotation of the crystal as possible and (2) the highest resolution data possible to better resolve the spatial coordinates of the electron density. As a reminder, high-resolution data in diffraction experiments are found closer to the edge of the detector.

In many cases, a native dataset will not be enough to solve a crystal structure. Native data allows for determining the intensities and positions of the diffraction spots; however, information on the relative phases of the waves that produced the spots is also needed if a structure is to be solved. This inconvenience is known as the phase problem and is another large bottleneck in X-ray crystallography. Fortunately, collection of anomalous dispersion datasets can reclaim the phasing information; for more information on anomalous dispersion, see Sect. 5.2.

While collecting anomalous data, resolution becomes far less important. Instead, the goal of these experiments is to get as much data as possible without ruining it through excessive X-ray damage. While this balance between quantity and damage is always at play in data collection, it is especially important when collecting anomalous diffraction data, as the damage will alter the spot intensities in ways that hide any useful anomalous signal.

4.2 Variables Associated with Data Collection

Now, let us circle back to the question of which crystals are the "best" ones for collection and what aspects of the data collection can impact these criteria. As stated above, resolution is often the most toted aspect of a crystal's diffraction; one would be hard pressed to find a structural paper that does not proudly brag about their "X-Å structure." And rightly so, as resolution can have major impacts on how confident one can be when it comes to ligand placement, side chain orientation, and—for the IBP crowd—the position of ice-like waters. However, to get higher resolution requires an increase in X-ray intensity, which goes hand in hand with an increase in X-ray damage. Also, the distance between the detector and the crystal (*see* Fig. 7, parameter d) must be optimized to keep high-resolution spots on the detector (move detector closer) without squishing the lower-resolution data together.

Beyond simple resolution, diffraction patterns hold a lot of information regarding the crystal lattice and potential pathologies within. Mosaicity is a measure of how well the lattice unit cells are packed together, with poorly packed crystals producing smeared diffraction spots that are difficult to process. Diffraction patterns may be complicated by overlapping lattices, either due to multiple crystals in one loop or multiple lattices in one crystal; while not ideal, many data processing programs are able to separate multiple lattices if other pathologies are absent. More problematic is twinning, a multi-lattice pathology in which a crystal is made up of symmetrically related lattices. Unlike multiple lattices that are randomly oriented and therefore produce separate (and distinguishable) patterns, the symmetrical relationship between twinned lattices means that select diffraction spots will overlap completely, meaning that the intensities will be a complex amalgam of two (or more) patterns. While there have been many computational techniques created to try and de-twin data, it is often easier to screen for a different crystal morphology that lacks this pathology.

If you are lucky enough to find multiple crystal morphologies for your protein, another variable to consider is the space group. The space group is indicative of the symmetry of the crystal and, therefore, how much data you need to collect to have a "complete" dataset. The higher the symmetry of the space group, the less data you need to collect. High symmetry space groups are therefore desirable, especially for anomalous scattering experiments where high completeness is needed while keeping X-ray damage low.

4.3 Choose Your Weapon: Synchrotron vs. Home-Source Diffractometers

Armed with the knowledge of what experiments you can do on what kinds of crystals, there remains a big question: where will you collect your data?

Select structural biology labs have home-source diffractometers that allow their members (and paying associates/collaborators) to collect data at home. These machines produce X-rays via bombarding certain metals, like copper, with accelerated electrons. The advantages of having a home source are obvious: the ability to screen crystals immediately, no need to wait to collect, no need for shipping your crystal elsewhere, etc. However, these diffractometers can only produce a single wavelength of X-rays, determined by the metal being used. While not a problem for native collection, most anomalous scattering experiments become impossible. Additionally, the low intensity of the X-rays means that the maximum resolution available is limited, even for otherwise strongly diffracting crystals. Throw in the pragmatic issues of upkeep and maintenance, and it begins to make sense why many labs are foregoing a home source and instead relying on synchrotrons.

Synchrotrons used for protein X-ray crystallography studies are cyclic particle accelerates that produce and store beams of high-velocity electrons. The electrons can be oscillated to produce high-intensity, tunable X-rays, which are then channeled into end stations (also known as beamlines) for their use in a myriad of scientific endeavors. There are many publicly funded synchrotrons around the world that accept proposals from incoming scientists. While in the past the only way to collect data at a synchrotron was on site (meaning expensive travel fares and stressful all-night collections in person), most synchrotrons now offer more convenient options. Mail-in collection is very helpful for new or infrequent users, as it only requires shipment of the crystals to the synchrotron where the collection will be handled by a beamline scientist. Meanwhile, remote access collection allows users to collect from the comfort of their own lab using a stable Internet connection and on-site robotic arms that move samples to the X-ray beam for collection.

The advantages and disadvantages of synchrotrons versus home-source diffractometers are inversely related; synchrotrons lose the convenience and immediacy but gain the X-ray brightness and multi-wavelength functionalities. The more structured nature of synchrotron time can be partially overcome through block agreements, where certain synchrotrons promise more regular beam time to communities (institution- or city-wide) of structural biologists. An added bonus is the beamline scientists that work at synchrotrons; their expertise is a valuable resource for figuring out the best way to run collection experiments on their particular beamline.

5 Structure Solution

After a long night of collecting data from a synchrotron, you are likely to have multiple datasets, each containing hundreds of frames that have accumulated into gigabytes of data. From this glut of images containing nothing but black spots on a mostly white background, a protein structure will (hopefully) be elucidated. The programs and processes used to move from spots to structures are numerous, with different labs and even different members of the same lab having their own preferred methods. Furthermore, many of the software packages offer tutorial datasets and walk-throughs that explain their processes much better than could be done here in simple text.

Therefore, this section will instead aim to explain the general pipeline for structure solution (*see* Fig. 8) while introducing the current software that is in use for these purposes.

5.1 Data Reduction

Having acquired the data, one must next reduce the dataset down from hundreds of image files into a single file that contains an ordered and processed inventory of all the spot-intensity information from the entire dataset. The most popular programs for easing this otherwise grueling task are XDS [19] and DIALS [20], though older programs like iMosflm and HKL2000 are still readily in use. At the end of this process, most programs will produce a reflections file with a ".mtz" extension (alternatively .hkl or .sca) that can then be used for further structure elucidation.

Data reduction (also referred to as data processing) contains several discrete processes involved in all forms of the pipeline. The first is indexing, which estimates the space group and unit cell parameters of the crystal. Indexing relies mainly on the relative position of the spots on the detector and therefore requires spot-finding, where each frame of the dataset is analyzed for points of increased intensity relative to the background (i.e., a diffraction

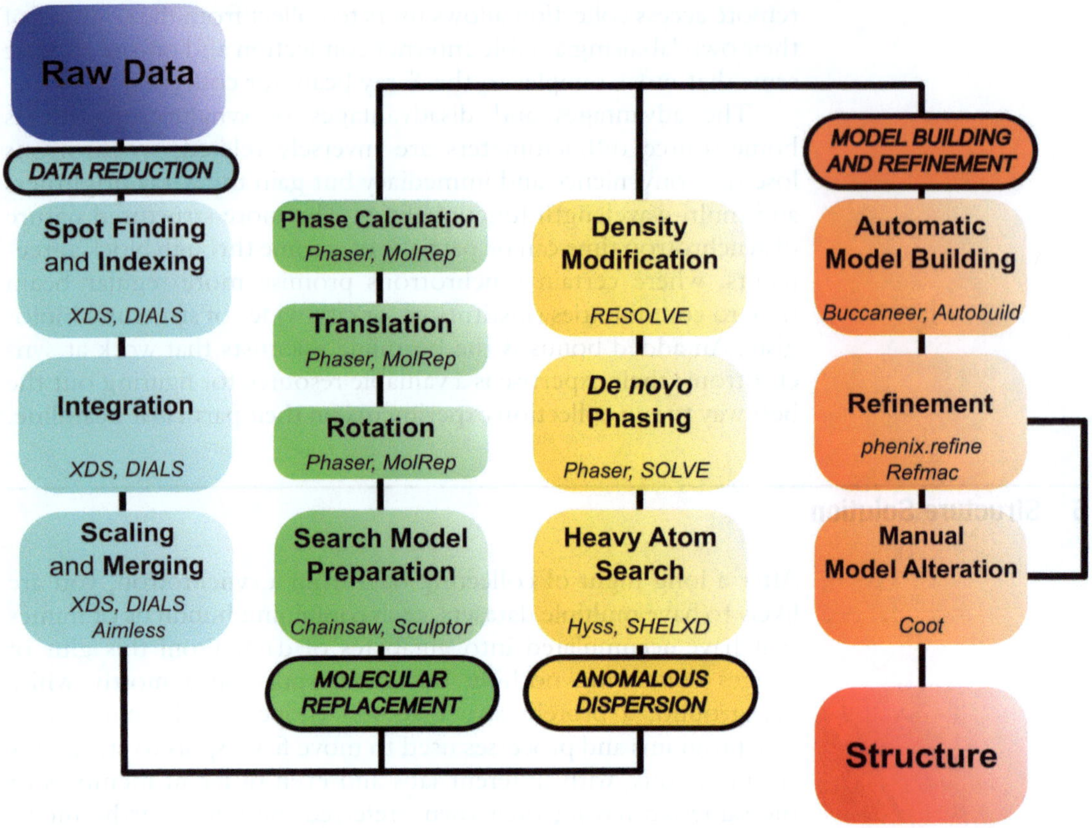

Fig. 8 Workflow for solving a structure. A general outline for the processes required to take diffraction data and turn it into a solved structure ready to deposit into the PDB. The programs listed in italics for each step are by no means an exhaustive list

spot). That said, a good estimate of the unit cell does not necessarily require the entire dataset; certain programs can index using as little as a single frame. The indexing quality is dependent on many factors, including the accuracy of the collection parameters such as detector distance, beam position, and X-ray wavelength. Post-index, several rounds of refinement to ensure the accuracy of these parameters are run before proceeding.

The second step in data reduction is integration, where the intensity of the spots for the entire dataset is measured and corrected for background intensity. There are two general methods of integration: 2-D and 3-D. The former is an older method that independently measures the spot intensities of each image and stores them for later processing. In other words, 2-D integration looks at the intensity of spot pixels only in the two dimensions available in a single image. 3-D integration is a newer method used by XDS and DIALS, which adds in the third dimension of between frames, amalgamating partial reflections from multiple frames and indexing them as a single full reflection.

The third and fourth portions of data reduction are scaling and merging, respectively. Scaling attempts to take imperfections in the collection process into account by placing the measured reflections onto a similar scale. In doing so, systematic changes in crystal diffraction (e.g., X-ray damage that reduces signal strength throughout collection, crystal dimensions that change how much crystal is interacting with the X-rays during rotation, changes in the intensity of the X-ray beam itself) can be mitigated. Merging is the combining of symmetry-related reflections together to remove redundancy.

The outputs from scaling/merging programs like Aimless [21] are a great tool for better understanding the quality of the data you have. Often times, these programs will help decide if the resolution of your data should be cut back, if there are poor frames within the dataset that can be omitted, or if there is an underlying pathology (e.g., twinning) that needs to be dealt with. A casual glance at an intensive crystallography publication will reveal the ever-daunting collection of statistics relating to both the quality of the collected data and the solved structure. Figure 9 highlights a few of the most

Table 1.

	Examplase Data
Synchrotron	APS 24ID-C (NE-CAT)
Wavelength (Å)	1.6314
Resolution range (Å)	43.44 –2.00 (2.072 –2.000)*
Space group	$P2_12_12_1$
Unit cell a, b, c (Å) α, β, γ (°)	27.6 44.5 195.9 90 90 90
Total reflections	57321 (3492)
Unique reflections	16683 (1503)
Multiplicity	3.4 (3.2)
Completeness (%)	96.7 (90.1)
Mean I/sigma(I)	8.6 (1.2)
Wilson B-factor	37.82
Rmerge (%)	6.6 (65.4)
Rmeas (%)	8.8 (91.8)
Rpim (%)	4.6 (48.5)
CC1/2	0.997 (0.675)
R-work (%)	20.1 (22.0)
R-free (%)	25.5 (30.1)
Number of non-hydrogen atoms	1811
macromolecules	1736
bromide	10
solvent	65
RMSD (bonds; Å)	0.005
RMSD (angles; °)	1.02
Ramachandran favored (%)	95.41
Ramachandran allowed (%)	4.59
Ramachandran outliers (%)	0.0
Clashscore	1.76
Average B-factor(Å²)	44.0
macromolecules	44.31
solvent	34.41

* data in brackets refers to only the highest resolution shell

Amount of data collected (completeness) and how many times that data was measured (multiplicity). **Maximize both.**

Signal-to-noise ratio **At high resolution: 1.0<**

Correlation coefficient **At high resolution: 0.3<**

Analysis of model's molecular geometry. **RMSD: low as possible** **Ramachandran outliers: <0.5%** **Clashscore: as close to 0 as possible.**

Indicators of precision. Rmerge: <15%, though considered obsolete **Rmeas overall: <15%** **Rpim overall: <10%**

Indication of how well model fits data. Value is partially reliant on resolution. **Should be as low as possible.**

Fig. 9 An example of a crystallographic "Table 1," with select parameters expanded upon. The two parameters in blue are most frequently used to determine where to cut data in terms of resolution

important parameters, with an attempt at relaying the conventional wisdom regarding what these parameters should be for "good quality" data.

5.2 Solving the Phase Problem

Armed with your .mtz file, the next step is to actually solve the structure. As one may expect, this is often easier said than done, as the phase problem now rears its ugly head. The .mtz file contains all the intensity information, which, once again, is related to the amplitudes of the waves that produced the diffraction pattern. To solve the structure, the lost phase information must be regained. There are many methods that can be used to do so, and most can be accomplished using the two program suites of Phenix [22] and/or ccp4 [23].

Two methods for solving structures that are rarely used these days are direct methods and multiple isomorphous replacement (MIR). The former is a statistical methodology of calculating phase information, which is only possible with very small proteins that diffract to very high resolution; therefore, this method is not applicable to most protein crystallography studies, though studies of small IBPs could find this useful [12]. The latter is an older method that adds heavy atoms—high-atomic-number elements that alter the X-ray diffraction—to crystals and calculates phase information by comparing datasets with and without these added atoms. However, MIR only works if the addition of heavy atoms—either via co-crystallization or soaking of the crystal in a solution of the heavy atoms—does not alter the crystal morphology; even small changes to the unit cell parameters can render this technique useless. Due to these difficulties, most crystallographers turn to either molecular replacement or anomalous dispersion for solving the phase problem.

Molecular replacement (MR) uses previously solved structures (referred to as search models) to estimate the phase information of unsolved structures. For this to work, the solved and unsolved proteins must be structurally similar—the closer the structural resemblance, the greater the likelihood of MR success. As a rule of thumb, search models with at least 40% sequence identity to the protein of interest are considered promising. Still, search models can be improved through the removal of extraneous sequences that are not present in the protein of interest or by pruning the model's side chains back to the β carbon. These methods of search model alteration can be done manually or through a myriad of different programs in both ccp4 (Chainsaw) and Phenix (Sculptor). If no search model with a significantly high sequence identity exists, homology models can also be successfully used for MR. Modeling software such as Phyre2, Robetta, iTasser, and AlphaFold are all of potential use in this endeavor.

Once you have an appropriate search model, MR programs (such as Phaser or MolRep) must identify where the equivalent portion of your protein resides within the crystal lattice before the phase information can be calculated. Users provide their reflections file, the search model, and the sequence of the protein of interest. Phaser or its equivalent will then calculate Patterson functions—maps of inter-atomic distances that do not require phase information—for both the search model and the diffraction data. Patterson functions are too complicated to be of much use for structure solution for large molecules, but they can work as mathematical signatures for structure positioning within a crystal lattice. Thus, Phaser can rotate and translate the search model within the crystal lattice, searching for the position in which the Patterson functions for both the search model and data overlap. Once the search model is properly placed, the phase information will be calculated and a potential solution provided. In Phaser, the LLG and Z scores (referred to as TFZ in Phenix) can help determine whether the solution is correct. According to manuals for both ccp4 and Phenix versions of Phaser, an LLG score should be positive and "as high as possible," while the Z score should be above 8 for definitive solutions (though above 4 may indicate a weak-but-potentially-correct solution).

Despite the best efforts of Phaser and its peers, MR is insufficient for being able to solve novel folds, such as the many novel IBP folds discovered throughout the years. In these cases, collection of anomalous dispersion data can provide the answer. The theory behind anomalous dispersion is intensive, requiring more space than is currently afforded within this section. As a quick summary, we return to Braggs law and its view of the crystal as a series of parallel planes. For every set of planes, there are two potential reflections produced depending on whether the light approaches from the top or the bottom of the planes. These equal-yet-opposite reflections are observed in diffraction data and are called Friedel pairs; both pairs will have the same spot intensity but inverse phases. The goal of an anomalous dispersion experiment is to incorporate atomic elements that produce non-identical Friedel pair intensities at certain X-ray wavelengths. The differences in pairs produced by these "anomalous scatterers" can then be used to calculate phase information.

Many heavy atoms can function as anomalous scatterers, either in their native single-atom state or as clusters that are sold for the express purpose of anomalous dispersion. Less hazardous alternatives are the halides bromine and iodine, as well as the intrinsically present sulfurs in amino acids cysteine and methionine, though this anomalous signal can be more difficult to detect. Alternatively, the metalloid selenium can be used to replace sulfur in methionine, providing stronger anomalous scatterers that are incorporated directly into the protein.

Different anomalous scatterers will produce anomalous signal at different X-ray wavelengths, hence why the tunability of synchrotrons is required for these experiments. To determine which wavelengths to use, most synchrotrons are outfitted to run automated X-ray fluorescence experiments, where the crystal is subjected to a spectrum of X-rays in search of the scatterer's absorption edge: the wavelength at which maximal Friedel pair changes can be observed. Next, data is collected at one to three wavelengths around the absorption edge. For single-wavelength anomalous dispersion (SAD) phasing experiments, a single wavelength very close to the absorption edge is used, while multiple-wavelength anomalous dispersion (MAD) phasing chooses wavelengths that are both close and removed from the absorption edge for comparison purposes. While MAD is better able to resolve ambiguities in phase information, SAD is often used as it is less strenuous and less likely to cause radiation damage, which (as noted in Sect. 4.1) is especially detrimental to anomalous dispersion experiments. Once collected, the multiple processes (*see* Fig. 8) involved in solving SAD/MAD data can be run through one of the several automated pipelines (e.g., Crank, AutoSol). These pipelines also allow for the inclusion of partial MR solutions to improve the phase information.

5.3 Model Building and Refinement

The immediate output from a successfully solved dataset is, sadly, not a full-fledged model of the exact atomic connectivity of the protein, but an electron density map: a three-dimensional plot that indicates where electrons are found in the crystal. When viewed through the program Coot [24] in the default settings, the electron density map will often appear as a series of twisting tunnels of blue mesh, with small blobs of green and red mesh spattered about. The blue mesh is called a $2F_0$-F_c map, and into its winding tunnels and protruding knobs can be built a model of the main chain and side chains, respectively. The green and red mesh are peaks in a difference map, often the F_0-F_c map, which indicates portions where the data and the model differ (green indicating electron density in the data that is missing in the model and red indicating electron density in the model that is not present in the actual data). While researchers can and do use these maps to manually build models, there are now programs that can provide a helping hand through automatically building as much of the model as possible. In fact, how easily programs like Autobuild [25] and Buccaneer [26] can build the model is a good indication of the quality of the solution.

Once at least a partial model has been built, the true "fun" of refinement begins. Refinement in crystallography is an iterative process that reconciles the observed electron density with the calculated electron density that would be produced by the model (*see* Fig. 8). This can be done by oscillating between manually changing the model in Coot to reduce spikes in the difference maps and using refinement programs that automate the global altering of the model to better fit the data. The main parameter used to assess the success of this process is the R-factor (also referred to as R_{work}), a measurement of the difference between the observed and calculated data shown as a percentage (*see* Fig. 9). For a perfect agreement between data and model, the R-factor would be 0%. Since there will never be perfect agreement, most R-factors for data approaching 2 Å resolution will fall below 25%.

Strangely, this process of refinement in crystallography does not just alter the model but also includes the altering of the data to better fit the model. While that seems almost sinful from a scientific perspective, it must be remembered that the electron density map has been constructed on phases that were calculated, not measured. Therefore, refinement programs also use the model to improve the calculated phase information, thereby improving the map so that the model can be better fitted so the phases can be further improved and so on and so on. The obvious danger here is the potential to over-fit the data to a model that is not representative of the actual structure. To prevent this calamity, crystallographers turn to the R_{free}-factor, which is an R-factor calculated on a small subset of the data that is not subjected to refinement. While the R_{free} will always be slightly higher than the R_{work}, it should still improve with iterative rounds of refinement as the model gets better at predicting the entirety of the data. If the R_{free} fails to improve or if the magnitude of its improvement is far less than the magnitude with which the R_{work} improves, the strategy being used for refinement may be over-fitting the data and should be avoided.

Another important factor that indicates the quality of the model during refinement is the molecular geometry of the proposed structure. Proteins, like all molecules, are limited in the angles their covalent bonds can explore; therefore, our models should follow these known restrictions in geometry. Ramachandran plots are a useful tool for finding residues that deviate from accepted phi and psi angles. Additionally, MolProbity [27] is a validation program that points out suspicious bond angles, bond lengths, or clashes between atoms that are too close together.

6 You Have a Structure. . .Now What?

After all that—the crystals, the diffraction, the blue mesh, the ever-changing iterations of Fig. 9—you finally have your structure. It is a fantastic feeling and a major accomplishment, but we are no longer living in the days where a structure, alone, can guarantee a high-impact publication. After all, the whole purpose of this trip down the yellow brick road of structural biology was to gain a better understanding of ice-binding proteins. So, now that you have this structure with all its i's properly dotted, its t's well crossed, and its prolines correctly isomerized, what are you going to do with it?

The first step is to check just how novel your fold is. As stated above, many IBPs have folds that are distinct from other known protein structures. It goes without saying that novelty in science is highly valued, and the discovery of new tertiary structures is worthy of note. The Dali server is a great tool that takes your refined structure model (as a .pdb file) and searches the PDB for proteins with a similar fold [28]. Dali's output provides a list of the structures that aligned with the provided query structure, as well as parameters to help gauge how well the structures overlap. Good hits will have low RMSD values, high LALI scores, and a Z score of 8 or above. Still, no table of nebulous numbers really replaces the human eye for determining if there is a useful overlap, so make sure to check the graphical representation of each alignment. Good hits with Dali can provide insight into the evolutionary past of the IBP (e.g., the type II AFP has a lectin-like fold [29]). Conversely, a paucity of good hits can indicate a novel structure and another example of IBPs adding to the repertoire of protein folds.

Solved structures can often be used to determine aspects of the protein's function. Fortunately, it is well known that IBPs use flat, hydrophobic surfaces to bind to ice. For many IBP structures, a quick glance at the surface representation of the protein in PyMOL can indicate which surface is likely responsible for ice binding. However, this is not always the case, especially for the globular examples from fishes [6] or the DUF3494 ice-binding proteins [11]. In these cases, a solved structure can inform mutagenesis studies, where potential ice-binding residues are mutated into larger, bulky residues that would disrupt the uniformity of a usual ice-binding surface. Alternatively, the ConSurf server can help identify functionally relevant surfaces by using sequence alignments between homologs to identify portions of your structure that have been conserved throughout evolution [30]; such studies work best with IBPs that are more widespread throughout the tree of life, such as the DUF3494 IBPs.

Structures can help identify more than just the site of ice binding. Since crystal structures are able to pinpoint the location of more ordered solvent molecules, solved structures have been used to visualize ice-like waters: waters pushed up against the hydrophobic ice-binding site that are a part of the anchored clathrate used to bind proteins to ice. The arrangement and spacing of these waters can even provide insight into which planes of ice the protein binds to and which residues are necessary for said binding. There is also a thriving community of IBP researchers that use solved structures in molecular dynamics (MD) experiments to better understand the kinetic and energetic parameters behind ice binding. While great care has to be taken with these studies to ensure that the results properly mimic real-world parameters, MD can help fill in the gaps that our static images of protein structures encased in crystalline lattices cannot.

And with that, we come to the end. Needless to say, this chapter is not an exhaustive resource on X-ray crystallography, much less on structural biology in general. There are hundreds upon thousands of resources available, from textbooks and published articles to long-running forums and tutorials on the subject. It is my hope that these few pages within the soup of available resources may serve as a nucleation point upon which new crystallographers can grow. After all, the fact that we can determine the structure of proteins is a thing of awe. An awe that is often followed swiftly by the age-old question: how? How can we determine the shape of a molecule so small that it is dwarfed by the wavelength of visible light? Well, now you know; what you do with it is up to you.

References

1. Sicheri F, Yang DS (1995) Ice-binding structure and mechanism of an antifreeze protein from winter flounder. Nature 375:427–431
2. Jia Z, DeLuca CI, Chao H, Davies PL (1996) Structural basis for the binding of a globular antifreeze protein to ice. Nature 384:285–288
3. Graether SP et al (2000) β-Helix structure and ice-binding properties of a hyperactive antifreeze protein from an insect. Nature 406: 325–328
4. Liou Y-C, Tocilj A, Davies PL, Jia Z (2000) Mimicry of ice structure by surface hydroxyls and water of a β-helix antifreeze protein. Nature 406:322–324
5. Leinala EK et al (2002) A beta-helical antifreeze protein isoform with increased activity. Structural and functional insights. J Biol Chem 277:33349–33352
6. Liu Y et al (2007) Structure and evolutionary origin of Ca^{2+}-dependent herring type II antifreeze protein. PLoS One 2:e548
7. Pentelute BL et al (2008) X-ray structure of snow flea antifreeze protein determined by racemic crystallization of synthetic protein enantiomers. J Am Chem Soc 130:9695–9701
8. Middleton AJ et al (2012) Antifreeze protein from freeze-tolerant grass has a beta-roll fold with an irregularly structured ice-binding site. J Mol Biol 416:713–724
9. Garnham CP, Campbell RL, Davies PL (2011) Anchored clathrate waters bind antifreeze proteins to ice. PNAS 108:7363–7367
10. Lee JH et al (2012) Structural basis for antifreeze activity of ice-binding protein from arctic yeast. J Biol Chem 287:11460–11468
11. Kondo H et al (2012) Ice-binding site of snow mold fungus antifreeze protein deviates from structural regularity and high conservation. PNAS 109:9360–9365
12. Hakim A et al (2013) Crystal structure of an insect antifreeze protein and its implications for ice binding. J Biol Chem 288:12295–12304
13. Sun T, Lin F-H, Campbell RL, Allingham JS, Davies PL (2014) An antifreeze protein folds with an interior network of more than 400 semi-clathrate waters. Science 343:795–798
14. Wang Y et al (2020) Carrot 'antifreeze' protein has an irregular ice-binding site that confers weak freezing point depression but strong inhibition of ice recrystallization. Biochem J 477: 2179–2192
15. Berman H, Henrick K, Nakamura H (2003) Announcing the worldwide protein data Bank. Nat Struct Mol Biol 10:980–980
16. Wukovitz SW, Yeates TO (1995) Why protein crystals favour some space-groups over others. Nat Struct Mol Biol 2:1062–1067
17. Niedzialkowska E et al (2016) Protein purification and crystallization artifacts: the tale usually not told. Protein Sci 25:720–733
18. Till M et al (2013) Improving the success rate of protein crystallization by random microseed matrix screening. J Vis Exp. https://doi.org/10.3791/50548
19. Kabsch W (2010) XDS. Acta Crystallogr D Biol Crystallogr 66:125–132
20. Winter G et al (2018) DIALS: implementation and evaluation of a new integration package. Acta Crystallogr D Struct Biol 74:85–97
21. Evans PR, Murshudov GN (2013) How good are my data and what is the resolution? Acta Crystallogr D Biol Crystallogr 69:1204–1214
22. Adams PD et al (2010) PHENIX: a comprehensive python-based system for macromolecular structure solution. Acta Crystallogr D Biol Crystallogr 66:213–221

23. Winn MD et al (2011) Overview of the CCP4 suite and current developments. Acta Crystallogr D Biol Crystallogr 67:235–242
24. Emsley P, Lohkamp B, Scott WG, Cowtan K (2010) Features and development of coot. Acta Crystallogr D Biol Crystallogr 66:486–501
25. Terwilliger TC et al (2008) Iterative model building, structure refinement and density modification with the PHENIX AutoBuild wizard. Acta Crystallogr D 64:61–69
26. Cowtan K (2006) The buccaneer software for automated model building. 1. Tracing protein chains. Acta Crystallogr D Biol Crystallogr 62: 1002–1011
27. Williams CJ et al (2018) MolProbity: more and better reference data for improved all-atom structure validation. Protein Sci 27:293–315
28. Holm L (2020) Using dali for protein structure comparison. Methods Mol Biol 2112:29–42
29. Ewart KV, Yang DS, Ananthanarayanan VS, Fletcher GL, Hew CL (1996) Ca^{2+}–dependent antifreeze proteins. Modulation of conformation and activity by divalent metal ions. J Biol Chem 271:16627–16632
30. Ashkenazy H et al (2016) ConSurf 2016: an improved methodology to estimate and visualize evolutionary conservation in macromolecules. Nucleic Acids Res 44:W344–W350

Chapter 4

Large-Scale Purification of Natural Ice-Binding Proteins by the Falling Water Ice Purification Method

Maya Bar Dolev, Chen Adar, Vera Sirotinskaya, and Ido Braslavsky

Abstract

We describe here a generic technique for purification of ice-binding proteins (IBPs) using a commercial ice machine. The method, which is called *falling water ice purification (FWIP)* [1], is based on the affinity of the proteins to ice, and it is therefore suitable for any IBP, natural and synthetic, with and without fused segments or domains. The FWIP method does not require the addition of tags to the protein, nor the use of resins and additives, and is suitable for large-scale purification. These features may turn FWIP useful also in the food and medical sectors.

Key words Ice-binding protein, Antifreeze protein, Ice affinity purification, Protein purification, Cold finger, Falling water ice purification

1 Introduction

There is a great potential in using ice-binding proteins (IBPs) in many fields involving freeze avoidance and freeze control. The frozen food industry and cryopreservation of cells, tissues, and organs are a few examples of such applicative directions. Yet, there is an evident gap between this potential and the actual in related research and development. A major challenge in introducing IBPs into industry is the high cost of their production. On the one hand, IBPs are available naturally in many organisms that are easy to grow and produce, like plants, fish, and insects. On the other hand, IBPs differ dramatically in size, structure, and molecular properties [2], and relevant extraction and purification methods for the untagged natural versions are limited and difficult to scale up. The lack of sufficient resources for high-quality protein hinders the development of practical technologies and limits basic research, in particular in food and cryopreservation [3].

Ran Drori and Corey A. Stevens (eds.), *Ice Binding Proteins: Methods and Protocols*, Methods in Molecular Biology, vol. 2730, https://doi.org/10.1007/978-1-0716-3503-2_4,

The affinity of IBPs to ice is an exceptional property that can be used to isolate IBPs not only from other proteins present in the organism but also from all other substances, including small molecules and nucleic acids. When the temperature of a growing ice in a dilute solution is lowered at a slow rate, ice grows slowly without finger growth. In this situation, water molecules are added to the ice and other molecules are excluded. IBPs and other (much less common) ice-binding substances are exceptions since their affinity to ice forces their incorporation into growing ice crystals. This principle is the basis of ice affinity purification methods [1, 4, 5]. While other methods are useful for medium protein quantities, the falling water ice affinity purification (FWIP) is developed in particular for large scale [1]. In this technique, the sample is applied into an ice machine designed for production of ice cubes in automatic cycles as long as the input reservoir contains water. Liquid from the reservoir is pumped to the top of metal plates and flows down on their surface back into the machine's reservoir. The temperature of the metal plates is gradually lowered, and ice begins to grow slowly. Since liquid from the reservoir is circulated, ice growth continues until the procedure is stopped. During ice growth, IBPs partition between the ice and liquid phases, while most of the other components remain in the liquid fraction. At the end of the process, the ice cubes are collected, and the unfrozen fraction (drain) is reused for a second cycle of purification. The procedure is presented in Fig. 1.

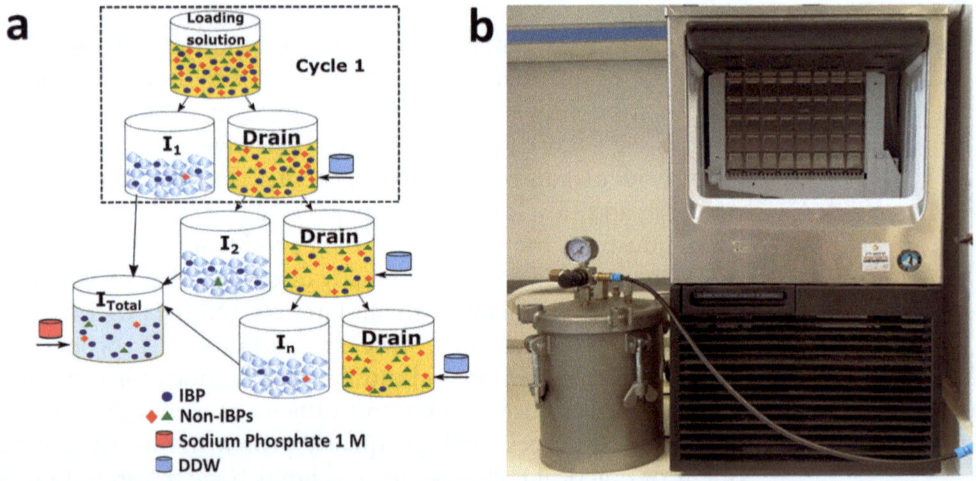

Fig. 1 FWIP: (**a**) The FWIP concept. A crude sample is loaded in the icemaker, and ice is produced. The unfrozen fraction is collected for reuse. This procedure is termed a purification *cycle*. The ice cubes are collected and melted together (I_1), and the drain fraction is diluted and loaded again for the next purification cycle. A series of such cycles is referred to as one purification *round*. The ice fractions from all cycles of FWIP are mixed (I_{Total}). If a higher purification level is required, this ice can be used as the starting solution for another purification round (not shown). Salt solution (small red tanks) is added to the ice fractions after melting to avoid protein aggregation. (**b**) The Hoshizaki KM-35A machine and the 10 L pressure tank used for FWIP. (Reprinted with permission [1])

The FWIP is an effective, low-cost, large-scale technique suitable for purification of any type of IBP, natural or recombinant, with or without tag. It does not require any resin, solvent, or additive and is therefore highly compatible with the requirements of the food and medical industries. FWIP does not produce any chemical waste, making it also environmentally safe. The purification efficiency can be increased by melting the purified ice and loading it again for a second round of purification. We obtained protein yield of up to 99.5% and purification level of 95% when purified IBP from crude lysates [1]. The volumes processed by the FWIP range from 0.6 L per cycle to hundreds of liters per day. These amounts are 100-fold larger than previous methods based on ice affinity purification, where solutions in the range of 0.1–1 L are typically processed in a few hours [4–6]. The processing of large volumes using FWIP allows production of grams of IBP per day, which can advance IBP research and the use of IBP in industry.

2 Materials and Special Equipment

2.1 Special Equipment

Ice machine (icemaker): The following protocol refers to Hoshizaki icemaker (Japan), model KM-35A (*see* **Note 1**).

Loading reservoir: A 10 L pressure tank (Kspark, Taiwan) (*see* Fig. 1).

Drain container: In the range of the inlet reservoir (10 L).

Scales: Such as T-Scale model SKW-L-300.

2.2 Materials

1. 1 M sodium phosphate buffer at pH 7.5 or 1 M sodium bicarbonate at pH 8 (*see* **Note 2**).
2. Double-distilled water (DDW).
3. IBP solution for purification.

Additional Requirements:

4. A method to concentrate the purified protein – such as the Ultrafiltration stirred cell (Merck, Germany) or a tangential flow filtration system (GE Healthcare, UK) equipped with a suitable membrane (*see* **Note 3**).
5. A method to determine total protein. We used the Bradford assay (Bio-Rad Protein Assay Dye Reagent Concentrate, Bio-Rad, USA).
6. A system for protein electrophoresis. We used sodium dodecyl sulfate-polyacrylamide gel electrophoresis (SDS-PAGE) with the PROTEAN electrophoresis system (Bio-Rad) following standard procedures [1].
7. A method to determine IBP concentration. We used a nanoliter osmometer (see Chap. 5 in this book) and SDS-PAGE.

3 Methods

3.1 Optimizing the Working Solution

Optimization of the starting conditions may significantly improve the protein yield and purification efficiency of each cycle and lead to a purified product in fewer FWIP cycles and rounds. We propose here a general procedure to optimize the starting conditions of the loading solution.

1. Estimate the total protein content of your sample using any standard technique such as BCA, Bradford, UV absorption, or SDS-PAGE.
2. If possible, estimate the fraction of IBPs in the total protein (*see* **Note 4**). This can be done by activity measurements or SDS-PAGE.
3. Calculate the total amount of impurities (all non-IBP proteins) in the solution:

$$\text{Total impurities [g]} = \text{solution volume [L]} \times (\text{total protein concentration [g/L]} - \text{IBP concentration [g/L]})$$

4. Dilute the IBP sample as follows:
   (A) The impurities (or total protein, if the IBP concentration is unknown or small) should not exceed 0.16 mg/mL (*see* **Note 5**).
   (B) The total salt concentration should not exceed 50 mM. We recommend working with 25 mM salts or less (*see* **Note 6**).
   (C) Avoid over-dilution as much as possible (*see* **Note 7**).
   (D) The minimal volume required for the ice machine is 0.6 L [1].

3.2 Adjusting the Ice Machine

1. Connect the inlet hose of the icemaker to the outlet of the pressure tank (*see* **Note 8**). We recommend a 10 L pressure tank; see Fig. 1b. For larger volumes, *see* **Note 8**.
2. Connect the silicone hose of the drain to the drain container. Fix the drain container below the water tank, since the drain runs by gravitation.

3.3 Running the FWIP First Purification Round

1. Place the pressure tank of the scales and auto zero the weight. Fill the pressure tank with the unpurified protein sample. Note the weight of your sample.
2. Adjust the pressure to 2 bar (*see* **Note 10**).
3. Plug in the power supply and move the control switch to "ICE" mode.

4. The icemaker will start automatically and pump the sample from the pressure tank into the icemaker's internal tank. Typically, 2.5 L is pumped (*see* **Note 11**).
5. *System initiation*: The icemaker starts an initiation procedure (sample pumping and primary defrosting stage), which takes a few minutes.
6. *Freezing stage*: After the initiation, the metal plates start to cool, and sample circulation begins. When the circulation begins, start a time count to estimane the duration of freezing stage(*see* **Note 12**). Typical duration for a 2.5 L starting solution is 40 min.
7. Note the scales. If all or most of the sample was pumped into the internal tank, i.e., the weight balance is close to zero (<0.7 Kg), refill the pressure tank with the drained solution and add 0.7 Kg of DDW. Pay attention to add the water during the freezing step, when the inlet valve is closed. There is no need to add water in case of dilute solutions (*see* **Note 13**).
8. Note the water circulation by sound. At the end of the freezing cycle, the water circulation stops, and the icemaker starts a defrosting stage (*see* **Note 14**). Stop the time count. This is the typical time of a freezing cycle of your sample .
9. *Defrosting stage*: While the ice cubes are still hanging on the cooling plates, dry-clean them briefly using airflow (*see* **Notes 15** and **16**).
10. Using a syringe or pipette, collect a small (50–500 μL) sample of the liquid fraction and a single cube of ice for analysis. Collect the remaining ice cubes from the ice bin.
11. At the end of the defrost step, a second cycle of freezing begins. The number of cycles should be determined empirically according to the yield and purification efficiency desired (*see* **Note 17**).
12. Collect the ice cubes for analysis and subsequent purification (another round; see Sect. 3.5) if needed.

3.4 Analysis of Purification Efficiency

1. Collect all the ice fractions from all cycles and mix together (*see* **Note 18**).
2. Melt the ice cubes and adjust the salt concentrations assuming the ice does not contain any salt (*see* **Note 19**).
3. Estimate the purification efficiency by SDS-PAGE of the melted ice, the drain sample, and the starting solution (*see* Fig. 2) (*see* **Note 20**).

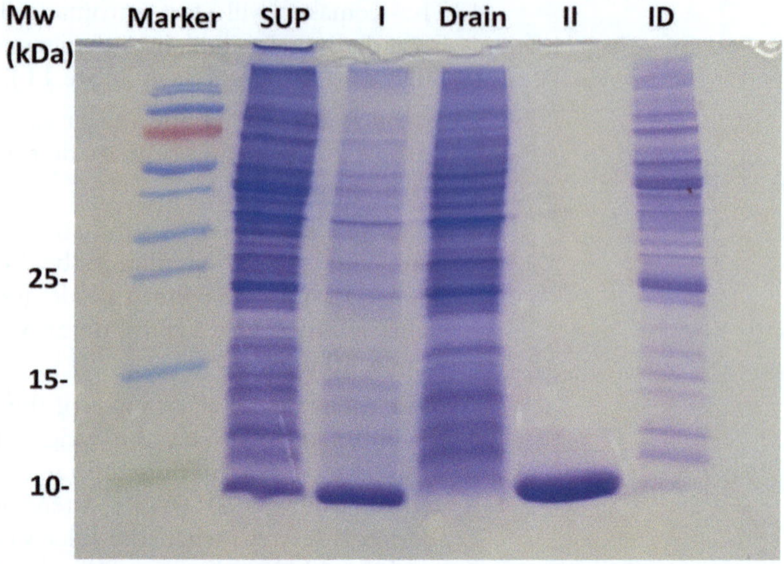

Fig. 2 SDS-PAGE analysis of type III AFP purified by FWIP. From left to right: marker (molecular weights in kDa are noted on the left); *E. coli* crude lysate supernatant before FWIP (SUP); combined ice fractions obtained from the first purification round (I); liquid fraction obtained from the first purification round (drain); combined ice fractions obtained from the second purification round (II); liquid fraction obtained from the second purification round (ID). The round II sample was concentrated twofold relative to the other samples, indicating >95% purity of the protein. (Reprinted with permission [1])

3.5 Second and Subsequent Purification Rounds

The purity of the IBP obtained can be increased by applying additional rounds of FWIP.

1. Wash the icemaker (see Sect. 3.6).
2. Collect all the ice fractions from all cycles and melt them together.
3. Adjust the salt concentrations (Sect. 3.4, **step 2**) (*see* **Note 21**).
4. Repeat Sects. 3.3 and 3.4 as needed.

3.6 Cleaning and Maintaining the Icemaker

Cleaning is recommended to avoid contaminations between runs. For a more extensive cleaning operation, refer to the icemaker manual.

1. Move the control switch to the "OFF" position.
2. Clean the pressure tank with water and fill it with freshwater or DDW.
3. Drain the water tank by disconnecting the silicone hose.
4. Reconnect the silicone hose back in its correct position after all of the water has drained.
5. Move the control switch to the "ICE" position to fill the water tank with water.

6. After 3 min, move the control switch to the "WASH" position.
7. After 5 min, move the control switch to the "OFF" position.
8. Repeat **steps 2** and **3**.

4 Notes

1. *Icemaker*: It is important to use an icemaker designed for producing transparent ice cubes.
2. Any buffer of choice can be used.
3. *If using a membrane-based concentration method*, the membrane cutoff should be considerably smaller than the molecular weight of the desired protein. We use a membrane with a molecular weight cutoff of 3000 for purification of AFPIII (9 kDa).
4. *Estimation of the IBP fraction* within the total protein is relevant for samples with a large amount of IBPs. For unknown samples, consider the amount of IBPs as negligible relative to the total protein.
5. *Impurities*: We found that when the crude sample contains up to 0.16 mg/mL of impurities, the purity of the product is optimal. When the crude sample consists of >0.3 mg/mL impurities, the amount of impurities in the purified product rises tenfold [1]. However, we note that too high dilution may lead to low protein yields.
6. *Salt concentration*: High concentration of solutes in the loading solution may reduce the efficiency of the purification due to nonspecific incorporation of molecules into the ice [5]. We therefore recommend low salt concentrations.
7. *Over-dilution*: Increasing the sample volumes will require larger loading volumes per cycle or more cycles. In either case, the purification will consume more time, and the purified protein will be diluted, which will then require more post-processing to increase concentration. In addition, working with highly diluted samples may lead to reduced protein yield and purification efficiency.
8. The Hoshizaki KM-35A icemaker is designed for producing pure ice from tap water by connecting its inlet pipe to a domestic water faucet. The machine is programmed for loading water until the internal water tank is full, and after every cycle of ice production, additional tap water is added to fill the tank. Some water is drained after every cycle, and after 10 cycles of ice production, all the contents of the water tank are flushed out. For IBP purification, we replace the inlet from the faucet to a pressure tank and the drain to an external container.

9. The Hoshizaki KM-35A can be connected to larger pressure tanks if needed. It can process up to 143 L in 24 h in automatic consecutive runs in a single operation. Larger versions of Hoshizaki ice machines are available (https://www.hoshizakiamerica.com), which could be used for processing volumes <143 L per day.
10. The pressure can be optimized empirically. We found that at pressures higher than 2 bar, bubbles are formed and interfere with the process.
11. Some of the pumped solution may be drained and can be processed in the next cycles.
12. Sample from the icemaker tank is pumped up to the top of the cooling plates and flows down on their surface, back into the icemaker water tank. During this sample circulation, the temperature of the cooling plates is gradually lowered, and ice cubes begin to form slowly. Consequently, the volume of the water in the icemaker tank is reduced. The time count of the freeze cycle is recommended for easier follow-up of subsequent cycles.
13. When the inlet tank is full, approximately 0.7 L of ice is produced each cycle. Therefore, 0.7 L should be added to the pressure tank after every cycle. However, if the weight balance is >0.7 L, meaning that a sufficiently large volume of sample is still in the pressure tank, addition of water is not necessary. Notably, addition of water is important when the sample gets concentrated with impurities and salts, but is not necessary for dilute samples. We recommend estimating the impurity level after every cycle and plan the purification strategy accordingly.
14. The beginning and termination of the freezing cycle are determined by internal controls of the icemaker.
15. The cooling metal plates are heated to melt off the ice cubes and release them into the ice bin. The duration of the heating is controlled by thermistor readings and usually takes a few minutes (up to 6 min). There is a short time window at the beginning of the defrosting stage, when the ice cubes still hang safely on the metal plates, for washing the ice from impurities without melting the cubes too much. For high purification yields, it is important not to miss this short time period before the ice cubes are released and another cycle of ice production begins. Note that at this stage, the inlet valve is opened and sample solution from the pressure tank is pumped into the internal water tank.
16. Sprinkling with cold water or a buffer is also possible, but might melt some of the ice.

17. In order to maximize the protein yield, more FWIP cycles are recommended. However, the IBP concentration in the drain is reduced every cycle, while the concentration of the impurities remains approximately unchanged. Therefore, after several cycles, the purification efficiency of FWIP drops.
18. Alternatively, use a single ice cube from each purification cycle.
19. For example, to get 25 mM of salt in a 1 L of ice, add 25.6 mL of 1 M buffer of choice. This is important for stabilizing the protein solution. Note that high salts should be avoided for subsequent purification rounds.
20. For efficient estimation, all samples should be concentrated by the same concentration fold. If the purification is insufficient, conduct another round of purification.
21. At this stage, the amount of impurities is expected to be lower than 0.16 mg/mL, so further dilution is usually not necessary, and the sample can be loaded into the pressure tank. However, if the amount of impurities is still high, follow Sect. 3.1.

Acknowledgments

We thank Elad Mor for helpful discussions and Tomer Friehmann for participating in the FWIP development. We acknowledge the Israel Science Foundation for financial support.

Conflict of Interest The authors declare no conflict of interest.

References

1. Adar C, Sirotinskaya V, Bar Dolev M et al (2018) Falling water ice affinity purification of ice-binding proteins. Sci Rep 8:11046.
2. Bar Dolev M, Braslavsky I, Davies PL (2016) Ice-binding proteins and their function. Annu Rev Biochem 85:515–542.
3. Nishimiya Y, Mie Y, Hirano Y et al (2008) Mass preparation and technological development of an antifreeze protein. Synthesiology Eng Ed 1: 7–14.
4. Marshall CJ, Basu K, Davies PL (2016) Ice-shell purification of ice-binding proteins. Cryobiology 72:258–263.
5. Kuiper MJ, Lankin C, Gauthier SY et al (2003) Purification of antifreeze proteins by adsorption to ice. Biochem Biophys Res Commun 300: 645–648.
6. Raymond JA, Fritsen CH (2001) Semipurification and ice recrystallization inhibition activity of ice-active substances associated with Antarctic photosynthetic organisms. Cryobiology 43:63–70.

Part II

Measuring and Quantifying the Activity of IBPs

Chapter 5

The Nanoliter Osmometer: Thermal Hysteresis Measurement

Nitsan Pariente, Maya Bar Dolev, and Ido Braslavsky

Abstract

The nanoliter osmometer is one of the most common tools in the study of ice-binding proteins (IBPs). It is used not only to measure the thermal hysteresis activity of IBPs but also to explore ice shaping, ice adhesion, and ice growth and melting rates and patterns. The advantage of the nanoliter osmometer for the IBP study and for studying single ice crystals lies in the small sample volume, in the range of nanoliters. Such a small volume enables precise determination and control of the temperature with precision in the range of millidegrees. This chapter describes in detail the process of determination of thermal hysteresis using a nanoliter osmometer operated by a LabVIEW interface. We describe the preparation of suitable capillaries and sample injection, which is a challenging step in the measurement. We then describe the procedure of single crystal formation and the determination of the melting and freezing temperatures. Insights on crucial parameters are emphasized.

Key words Thermal hysteresis, Thermal hysteresis gap, Single ice crystal, Nanoliter osmometer, Ice shaping, Crystal growth, Ice-binding proteins, Antifreeze proteins

1 Introduction

The phenomenon of thermal hysteresis (TH), which is the difference between the temperature at which ice crystals grow (the freezing temperature, T_f) and the temperature at which they melt (the melting point, T_m), has been a gold standard for characterization and quantification of the activity of ice-binding proteins (IBPs). In fact, IBPs were previously named thermal hysteresis proteins for their TH activity or antifreeze proteins for their ability to depress the growth of ice crystals at temperatures below the equilibrium melting temperature. IBPs were also classified as hyperactive or moderate according to their TH activity [1].

There are several methods to determine TH, including differential scanning calorimetry [2–4], sonocrystallization [5, 6], and nanoliter cryoscopy using a nanoliter osmometer [7–10]. The latter has been one of the main tools in the field, not only for measuring

Ran Drori and Corey A. Stevens (eds.), *Ice Binding Proteins: Methods and Protocols*, Methods in Molecular Biology, vol. 2730,
https://doi.org/10.1007/978-1-0716-3503-2_5,

TH but also for characterizing protein properties that affect TH, such as crystal shaping [11] and binding kinetics [12]. Using the nanoliter osmometer, ice crystals, typically 30 μm wide, are observed directly using optical microscopy. The crystals are grown in a drop formed by injecting a few nanoliter volume samples in oil-filled copper holders that are positioned in a cold stage with controlled temperature. A detailed description of the system has been published [7], and here we further extend on the method. The effective temperature controller, the small copper holder, the oil surrounding the sample from all directions, and the small sample volume allow fast response time and small temperature fluctuations, of only millidegrees. These features are particularly important to accurately determine both the T_m and the T_f, especially in the cases of low TH. The temperature fluctuation is in the range of 0.002 °C [13], which allows measuring extremely low TH activity, for example, in IBP mutants [14]. A range of properties related to interactions of substances and ice have been studied using the nanoliter osmometer: ice shaping [11], ice growth rates in cases where there is no complete growth arrest by the sample [15], melting hysteresis [16], and recrystallization inhibition studies [17]. In the case of the Antarctic bacterium *Marinomonas primoryensis*, we used the nanoliter osmometer to observe and characterize the interactions of the living and swimming bacteria when they encounter ice crystals in vivo [18].

Early designs of the nanoliter osmometer were developed already in the 1950 in order to determine the melting point depression of body fluids in nanoliter volumes [9, 19]. By 1973, Frick and Sauer constructed an improved version of a nanoliter osmometer for measuring the freezing point depression of insect hemolymph. They reported standard deviation of 0.05 °C between ten independent measurements [8]. Manually controlled nanoliter osmometers were manufactured by *Clifton Technical Physics* (Hartford, NY) [10] and by *Otago* (Dunedin, New Zealand) [20] and used to measure the osmolarity of biological samples in eye fluid such as tears [21] and later on in the IBP research [22, 23].

Based on the physical design of the *Clifton* nanoliter osmometer, Braslavsky and his team engineered a computer-controlled nanoliter osmometer with an automated temperature controller and high-quality recording options that broaden the experimental possibilities of the system. A temperature controller reduces the temperature fluctuations in the system, and the custom LabVIEW control platform (National Instruments) provides a range of digital features that significantly improve the convenience and experimental freedom of the system. These features allow determination of temperature with a resolution of millidegrees [7]. *μIce* implemented some more software and hardware improvements and offers a similar nanoliter osmometer commercially [24]. Figure 1

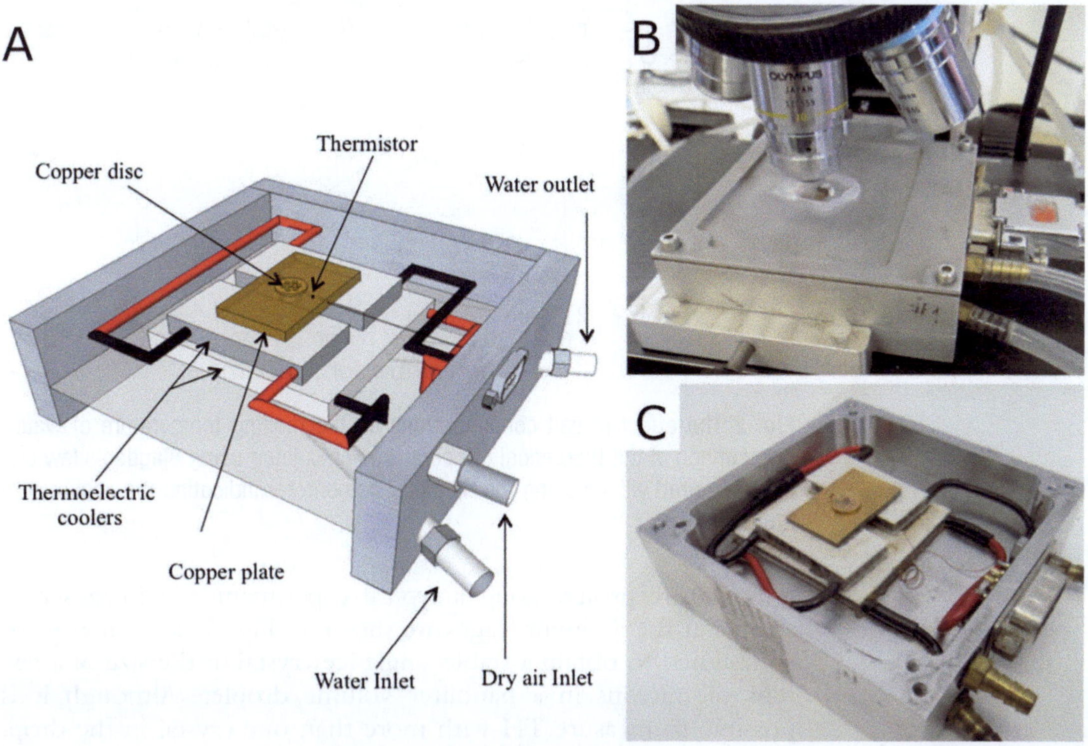

Fig. 1 The cold stage of the nanoliter osmometer. (**a**). A schematic presentation of the compartments inside the cold stage. (**b**). A picture of the cold stage when it is closed and ready for work. (**c**). A picture of the cold stage when it is open. (Reprinted with permission [7])

demonstrates the interior of the unique cold stage of a controlled nanoliter osmometer. To characterize the reliability of the T_m measurements using our osmometer and to assess the linearity of our calibration methodology, we measured the T_m values of a set of known salt solutions and compared them with the calculated correlation between T_m and osmolality following Blagden's law [25]. Figure 2 presents the measured values next to the calculated curve. Each point represents the average of four independent measurements. The error bars are smaller than the mark size. The small difference of only 0.03 °C between independent experiments indicates the stability and robustness of the osmometer, and the overlap between the measured and theoretical values indicates the reliability of the temperature calibration in the regime of 0 °C. The slight shift of the average value at one osmolar might be due to inherent nonlinearity, which becomes more dominant as the salt concentrations and supercooling increase, as previously noted [26].

The protocol provided here describes a typical experiment for determination of the TH gap of an IBP sample using the LabVIEW-controlled nanoliter osmometer. A presentation of the

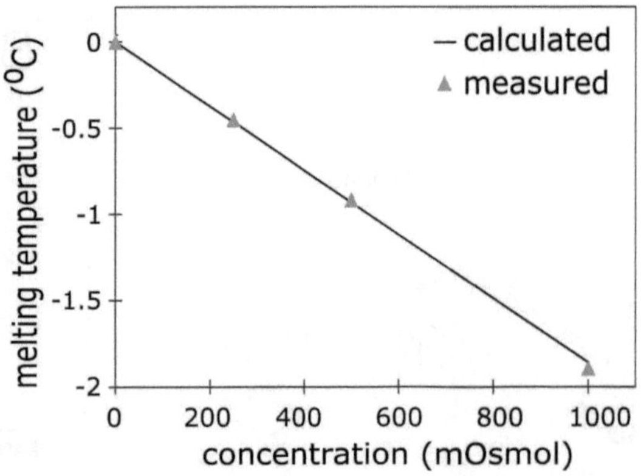

Fig. 2 The effect of salt concentrations on the melting temperature of water. Comparison of the theoretical temperatures calculated using Blagden's law and the measured values using the nanoliter osmometer indicating the accuracy of the apparatus

temperature profile along a typical experiment and images of a sample at the different stages are shown in Fig. 3. The experiment is designed to obtain a stable single ice crystal in the size of a few tens of microns in a nanoliter volume droplet. Although it is possible to measure TH with more than one crystal in the drop, the presence of only a single ice crystal is important to obtain the most accurate results. More than one crystal will increase the volume of the ice in the drop, which will increase the salt concentration and affect the colligative T_m. In addition, when more than one crystal is present, each crystal within the drop actually experiences a slightly different T_m due to small gradients within the drop. Differences in the crystal sizes, shapes, and defects on their surfaces will lead to different T_f, and the measured value will always be that of the first crystal that bursts. We therefore highly recommend having single crystals for proper TH measurement.

To obtain a single ice crystal, we first freeze the whole sample and then melt it back until a single ice crystal remains. Typically, freezing occurs at ~ − 35 °C, close to homogeneous nucleation, due to the small sample volume that lacks effective ice nucleators. While some IBPs are effective nucleators, small AFPs were found to only slightly influence the homogeneous nucleation temperature [27–29]. We then measure the T_m and T_f of this single crystal. The difference between these two temperatures is defined as the TH gap, as illustrated in Fig. 3. We also included a detailed description of sample preparation and injection, which may be challenging for an inexperienced user.

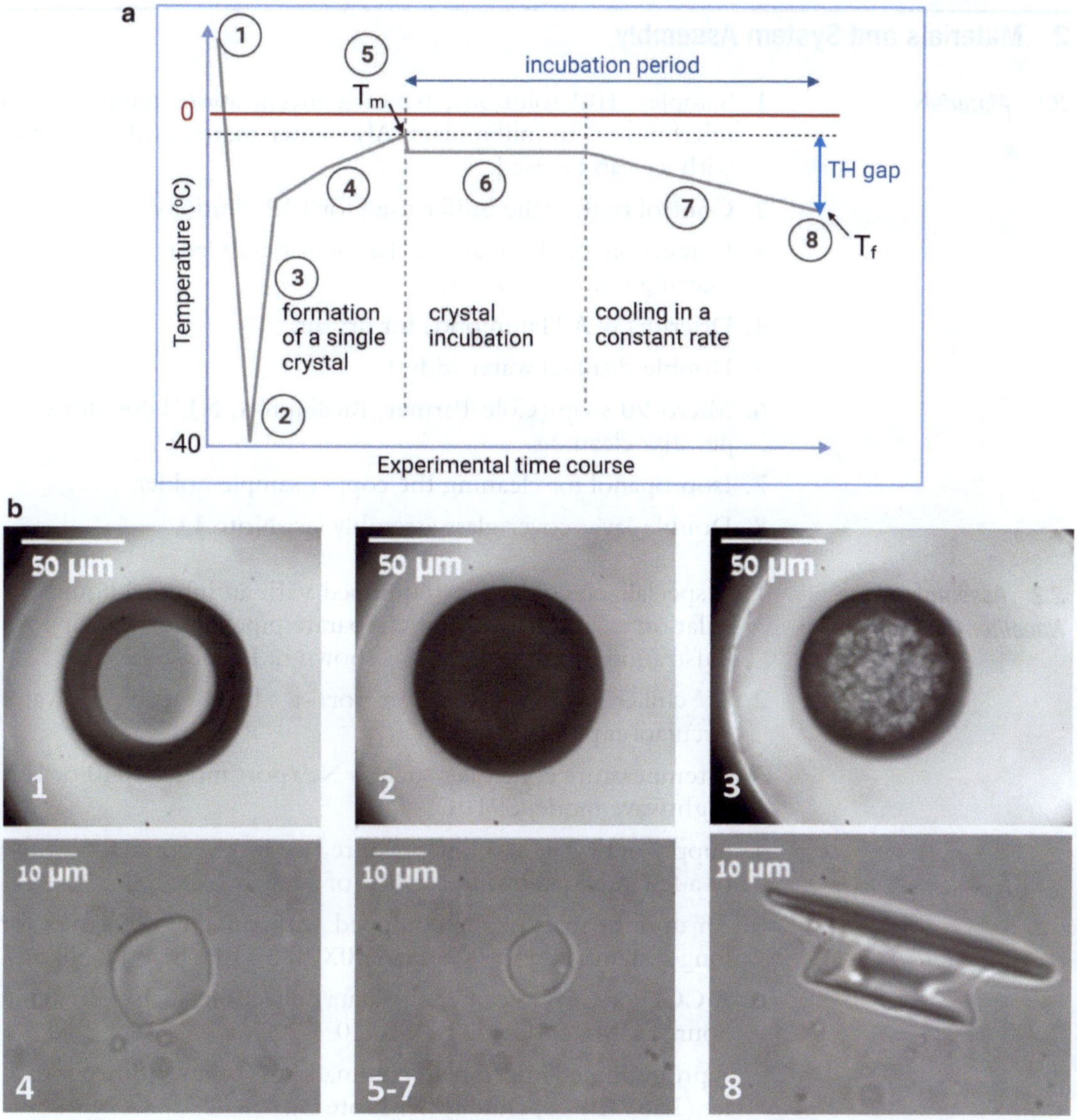

Fig. 3 The course of thermal hysteresis measurement experiment. (**a**) Temperature profile at the different stages of the experiment. (1) A sample at room temperature is transparent and lucid. (2) The sample is frozen at a fast-cooling rate. Freezing occurs typically below −30 °C. Due to the polycrystalline ice, light cannot pass and the drop appears dark. (3) The temperature is raised at a high warming rate and the ice is melted. The sample turns lucid again. (4) The warming rate is lowered as the temperature gets closer to the melting point and most of the ice is melted, until a single ice crystal remains. The crystal is shrunken to 20–30 μm diameter. (5) The melting temperature is determined by observing the lowest temperature at which melting is observed. (6) The crystal is incubated at a temperature slightly below T_m. Reshaping of the crystal may occur. (7) After incubation, the temperature is lowered at a constant rate and the crystal is recorded. During **steps 6–7**, the crystal size and shape remains constant. (8) The crystal bursts and ice fills the sample drop. The highest temperature at which growth is observed is the T_f. Created with BioRender.com. (**b**) Images of a TH experiment at the corresponding stages. In this experiment, 2 μM of *Tenebrio molitor* IBP fused to maltose-binding protein was used. The TH measured was 0.14

2 Materials and System Assembly

2.1 Materials

1. Samples: IBP solutions, typical concentrations ranging from micromolars to millimolars. Also other materials that interact with ice can be used.
2. Control buffer (the buffer used for IBP dilutions).
3. Immersion oil B (Cargille Laboratories, Cedar Grove, NJ), freezing point < −13 °C.
4. Drierite (W A Hammond) for dry air.
5. Double-distilled water (ddw).
6. Micro-90 soap (Cole-Parmer, Burlington, NJ, USA) for copper disk cleaning.
7. Isopropanol for cleaning the copper sample holder.
8. Double-layer cover glass assembly (*see* **Note 1**).

2.2 Assembly of the Nanoliter Osmometer

1. A specialized cold stage (equipped with an internal liquid circulation as a heat sink and a separate pipeline for dry air). An illustration of the apparatus is shown in Fig. 1.
2. A chiller with circulation or a bath with a water circulation pump.
3. A temperature controller such as Newport model 3040 or ILX Lightwave model 5910C.
4. Supply of dry air at a low pressure (we found optimal pressure of a few mbar, allowing air flow of a few mL/sec).
5. An upright microscope equipped with suitable objectives for long distance (such as CF Plan 50X/0.55 EPI ELWD, Nikon).
6. A CCD or CMOS camera for data collection, such as Imaging Source DMK 23UV024 USB 3.0.
7. A programmed interface to command the temperature controller (LabVIEW operated) (*see* **Note 2**).
8. A copper disk (sample holder) as shown in Fig. 1 with the following dimensions: 7 mm diameter, 0.5 mm thickness, 0.5 mm diameter cavities spaced 1 mm apart, center to center.
9. Injection needles: We use micro-hematocrit tubes (75 ± 0.5 mm in length, 1.15 ± 0.05 mm internal diameter) that were pulled with a capillary puller.
10. Glass syringe (10 mL, All Glass Luer-Tip Syringe).
11. Tubing to connect the water pump and the air flow to the stage as well as the syringe to the capillary.
12. Capillary puller (such as Narishige Model PB-7 Microscope Micropipette Puller).
13. A sonicator bath for the sample holder (copper disk) cleaning.

3 Methods

In order to prevent cross-contamination, it is essential to thoroughly clean the copper disks after every use. Each hole should be used only once (*see* **Note 3**).

3.1 Copper Disk Cleaning

1. Place the used disks in a glass beaker and immerse them in 20–40 mL of diluted Micro-90 soap (1:1000).
2. Sonicate for 10 min at room temperature.
3. Carefully discard the solution and wash twice with ddw to remove residual soap.
4. Repeat sonication in ~30 mL isopropanol.
5. Dry the disks using pressured dry air.
6. Using clean tweezers, collect the copper disks for storage in a clean and closed box or petri dish until use.

3.2 Glass Capillary Preparation

The procedure is presented in Fig. 4, **steps 1–3**.

1. A typical droplet used in the osmometer, 100–200 μm in diameter, contains few nanoliters of sample. We use thin capillaries to inject such small volumes to the center of oil-filled

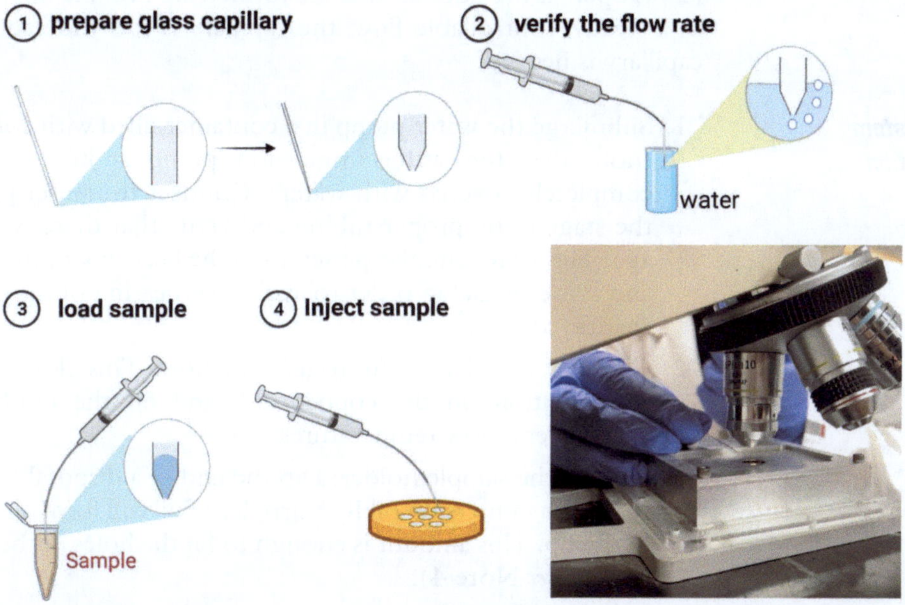

Fig. 4 Sample preparation for nanoliter osmometer. **Step 1**: A glass capillary is made using a pipette puller. The tip of the capillary should be open and narrow. **Step 2**: The flow rate of the capillary is evaluated by air bubbling into a tube of pure water. The firm bubbles formed indicate that the capillary width is suitable for sample injection. **Step 3**: Sample is loaded into the tip of the capillary. **Step 4**: Sample is carefully injected into one of the holes in the sample holder. An image showing the palm position during injection is shown on the right. Created with BioRender.com

cavities in the copper disk. One method to obtain sharp and narrow capillaries is to use a pipette puller. Suitable parameters for the pipette pulling depend on the pipette type, the puller type, and the external conditions and should be determined empirically. It is important to produce narrow capillary since the flow rate of injection is significantly influenced by the capillary width.

2. After the capillary pulling, the capillary obtained might be closed at the end. The capillary tip can be opened by cutting or breaking the tip, after which the opening might be too wide. For verification of a proper capillary opening, we test the capillary by its ability to form fine air bubbles in clean water before filling the capillary with the sample.

First, fill the plunger of the 10 mL glass syringe with a few mL of air. Then, connect the syringe to the capillary pipette using a short connector tube and dip the tip of the capillary into a 50 mL tube filled with water (*see* Fig. 4, **step 2**). Press the plunger to increase the pressure in the syringe to form air bubbles. If air bubbles do not form when strongly pressing the plunger, the capillary is closed, and there is a need to open it. Opening of the capillary can be achieved by gently touching the walls of the 50 mL tube with the sharp end of the capillary. The bubble test should be repeated to check if the capillary has been opened. Once the capillary is open, if the bubbles that are formed are not fine one-by-one, in a steady, controllable flow, the capillary is too wide and a new capillary is needed.

3.3 System Preparation

1. Submerge the water pump in a container filled with cold water (note that the water immersion pump should always be completely covered with water). Connect the water pump to the stage using proper tubing and verify that there is no leakage. This water circulation serves as the heat sink of the system, and it is essential in order to avoid overheating of the thermoelectric coolers.
2. Connect the dry air/nitrogen gas flow. This flow prevents condensation on the copper disk and on the double-layer glass cover at low temperatures.
3. Prepare the sample holder: Dip the end of a disposable tip in a tube with immersion oil B. A droplet of oil will form on the side of the tip. This amount is enough to fill the holes of the sample holder (*see* **Note 4**).
4. Using tweezers, invert the disk and place the sample holder in the cold stage under the microscope. Make sure the holes are clearly visible.

3.4 Sample Injection

1. Injection process can be observed directly through the binocular or by the camera real-time video on the computer screen. It may be more convenient to use the 10× lens for injection. Turn on microscope illumination, camera, and LabVIEW controlling program.
2. Pull a small amount of sample into the sharp end of the capillary from a sample tube using the glass syringe. Be careful not to pull air into the capillary after pulling the sample when the tip is not immersed in the sample tube. After loading the sample into the capillary, the sample should be inserted immediately to the oil in the holes of the copper disk to avoid change in concentration due to evaporation.
3. Stabilize your palm on the stage for injection (*see* Fig. 4). Direct the pipette tip to the center of one oil-filled cavity. Inject a small amount of sample into the middle of the cavity. Note that the drop is not shrinking, drifting to the edge, or touching the metal (*see* **Note 5**).
4. Cover the disk with a double-layer cover glass. The sample is ready for measurement.

3.5 Obtaining a Single Ice Crystal in the Droplet

In order to obtain ice crystals in a nanoliter volume sample, we need to nucleate the sample. Due to the small size of the droplet, spontaneous freezing occurs at temperatures < −30 °C. Therefore, we need first to freeze the whole sample and then to melt it back until only one ice crystal remains. The freezing is done by setting the temperature to low-temperature, fast-cooling mode. Once the sample is frozen, as detected by the darkening of the drop, a melting process may begin. We then melt the sample at a slower rate. As the temperature gets closer to the melting point, we use smaller temperature intervals for elevating the temperature. This is important in order to avoid overheating, which can melt the crystal completely, and for accurate determination of the T_m.

1. Press the "cooling on" button. The controller's TEC light will light up. This turns on the fast-cooling program, which sets the rate of temperature change module to the fastest possible and sets temperature set point to −40 °C. The temperature will drop rapidly.
2. Wait for the sample drop to freeze while following the temperature decrease. The sample will freeze between −30 °C and −40 °C, depending on the ionic strength of the solution and the presence of nucleating substances. When freezing occurs, the sample will abruptly turn dark due to the fast formation of dendritic ice that may further darken due to recrystallization that may occur in the drop.
3. Once the drop has frozen, set the temperature to −10 °C.

4. Raise the temperature gradually until significant parts of the sample are melted and single ice crystals can be observed. This is a good time to switch to the 50× objective.
5. Continue to raise the temperature slowly. As more crystals melt, raise the temperature in smaller steps, which keeps the temperature gradients small and prevents overheating. We find the step buttons best for the fine temperature changes necessary when only a few crystals remain and most of the sample has melted. Use steps as low as 0.002 °C. Continue the process until a single ice crystal remains. If the crystal is big, continue to melt it to a size of ~20–30 μm.

3.6 Determine the TH

1. **Determine the melting point of the crystal**: The melting point of the crystal is defined as the highest temperature at which melting was observed.
2. **Determine the freezing point of the crystal.**

   The freezing point is defined as the highest temperature at which the ice crystal begins to grow. The growth rate depends on the supercooling at the freezing point. At high TH activities, i.e., low freezing points and high supercooling, the growth will be fast and is referred to as "burst." Typical growth patterns of ice in the presence of a moderate IBP from fish and a hyperactive IBP from an insect are shown in Fig. 5. The differences in growth patterns are due to different affinities of IBPs to various ice planes. Before lowering the temperature, we allow accumulation of protein on the crystal surface for a

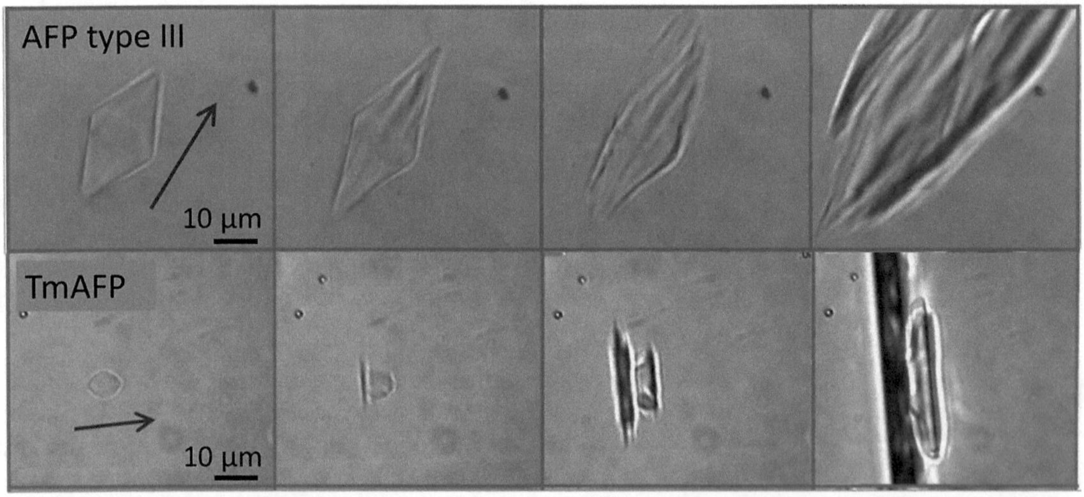

Fig. 5 Ice growth ("burst") patterns in different IBP solutions. The sequence of images shows the characteristic ice shapes close to the freezing point before the "burst" (left frame) and immediately after the freezing point is reached (the following frames). The arrow indicates the direction of the crystallographic c-axis of the hexagonal ice crystal. Top: solution of 200 μM of eelpout AFP. Bottom: solution of 5 μM insect IBP from *Tenebrio molitor*. (Adapted with permission (11))

constant period of time (*see* **Note 6**). Within the thermal hysteresis gap, reshaping of the ice crystal may occur, depending on the IBP type [11].

3. **Incubation and cooling**: Set the temperature to slightly below the T_m in order to prevent the crystal from melting. Let the crystal rest for a defined time period using the "delay" option. Typical annealing conditions are 10 min at −0.1 °C below T_m (*see* **Note 7**). Set the desired temperature ramp and start recording. For optimal determination of the cooling rate, *see* **Note 8**. The cooling rate can be changed during the course of an experiment by turning off the ramp button, changing the parameters, and pressing the ramp button again. The experiment is complete when sudden growth of the crystal occurs. The temperature at this instant defines T_f.

3.7 Calculate the Thermal Hysteresis Gap

Thermal hysteresis will be calculated as $TH = T_m - T_f$. We recommend repeating the experiment at least three times with different droples for each concentration.

Figures 6 and 7 demonstrate the TH as a function of experimental conditions and protein concentrations. The TH typically correlates linearly with the square root of the protein concentration [6].

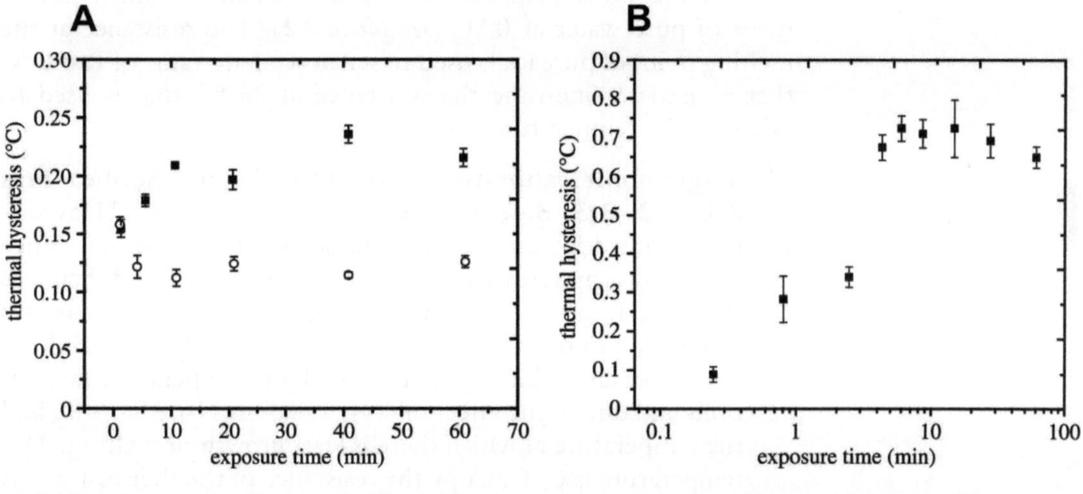

Fig. 6 Effects of incubation time and temperature on thermal hysteresis values for two different IBPs. (**a**) Fish AFP III (40 mM). TH values are the average of at least three measurements taken at 0.028 °C below T_m (open circles) and 0.18 °C below T_m (filled squares), with the variability indicated by the vertical error bars. (**b**) IBP from the Antarctic bacterium *Marinomonas primoryensis* at 2.4 mM. The TH values are plotted against the log of the exposure time. The shown values are the average of at least three measurements with the variability indicated by the vertical error bars. (Adapted with permission from (12))

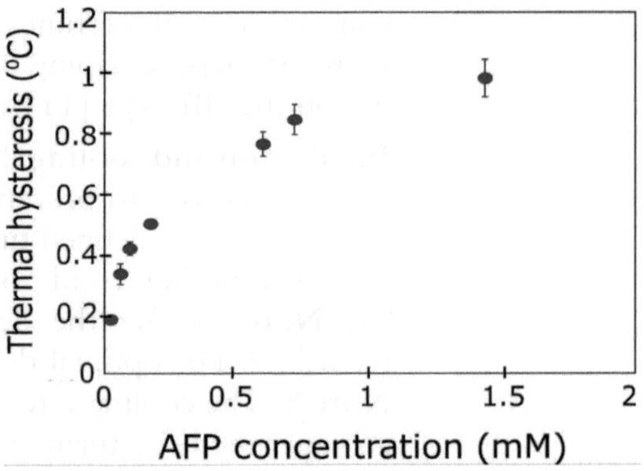

Fig. 7 TH of fish AFP III as a function of concentration, measured following the protocol described above. Each value represents the average of six to nine independent measurements. Error bars represent the standard deviation

3.8 Melting Hysteresis

While the TH is a dominant effect of IBPs, it was found that they also have some melting inhibition activity. A detailed description of melting hysteresis experiment has been published [16].

3.9 Temperature Calibration

The calibration of the osmometer is composed of two steps. The first step is permanent and includes several parameters that are set by the manufacturer, with a free parameter that is the resistance of the thermistor at 25 °C (*see* **Notes 10 and 11**).

The second step of the calibration is a single-point measurement of pure water at 0 °C (*see* **Note 12**). The resistance at the melting point of pure ice is used to set an accurate value of R at 0 °C that is used to fine-tune the resistance at 25 °C that is used to calculate the temperature from the resistance measurement.

1. **Single-point calibration procedure**. Follow **Subheadings 3.1, 3.2, 3.3, 3.4, 3.5, and 3.6** for a regular TH experiment with ddw, until a small single crystal is observed. Since there is no hysteresis in pure water, the crystal is difficult to stabilize. It grows when the temperature drops and shrinks when the temperature is raised. It is practically impossible to have a totally stable crystal due to slight temperature fluctuations and due to the effect of crystal size (*see* **Note 9**). Still, find the temperature at which there is least growth or melting. This temperature is 0 °C. Copy the resistance of the thermistor ("R actual") at T = 0 °C value to "R(100 μA)(T = 0)." This sets the zero point of the osmometer for the 100 μA mode. It is recommended to also calibrate the R(10 μA) for the 10 μA mode (*see* **Note 10**).

4 Notes

1. **Cover glass assembly**. After injecting of the proteins to the sample holder, it is essential to cover the opening of the osmometer stage to prevent moisture from accessing the sample. The cover should be transparent and thermally isolated to avoid accumulation of humidity and ensure optimal light transmission. We found that a single-layer cover glass does not provide sufficient thermal insulation. We recommend using a double-layer cover glass by adhering two cover glasses with 1–3 mm spacing and placing a Drierite particle (1 mm in diameter) in the space between them. Simple wax, nail polish, or hot glue are suitable to seal some air between the glasses.
2. **LabVIEW subroutine**. The subroutines developed at the Braslavsky Lab can be obtained by request from the corresponding author. Further details are given in [7]. In addition, the company *μice* has developed a similar LabVIEW subroutine that is included in their nanoliter osmometer [24].
3. **Copper disk cleaning**. Many IBPs are stable and can be restored after long storage in dry state. Residual samples in the cavities of the sample holder can then lead to contamination. We recommend washing the sample holder after each use of all the sample holes.
4. **Filling the sample holes with oil**. All of the sample holes should be firmly filled with the minimal amount of oil. This way, the light passing through the holes is optimal for observation. If there is too much oil, a heavy drop with high curvature will form, and the holes will appear dark. The user can use the holes that are not dark or replace the sample holder.

   Another possibility is to use two types of oils, one with density lower than water and the other with higher density. This way, the aqueous drop would reside between the two oil types, reducing thermal gradients within the oil [30].
5. **Sample injection**. The sample should be centered as much as possible in the oil and remain stable over time. If the sample touches the copper walls, a significant temperature gradient will form in the droplet. This may lead to inaccuracies in temperature determination and in many cases to evaporation of the drop during experiment. Alternative to injection of the droplet by hand is injection by three-axis micromanipulator.
6. **Crystal incubation time and temperature**. Different types of IBPs have different accumulation kinetics, which may affect the freezing point of the crystals and the overall TH by several folds. This effect is presented in Fig. 6. It is therefore important to keep the exposure time and temperature interval constant in

every set of experiments. Note that during cooling, the accumulation continues, and therefore, the cooling rate and the time that passes until the T_f is reached also influence the accumulation time and need to be considered. For consistency, we recommend a minimum of 10 min dwell before the temperature ramp down starts.

7. **Factors that influence the TH gap**. The TH gap is influenced not only by the incubation time (*see* Fig. 6 and **Note 6**) but also by the type of the IBP, its concentration (*see* Fig. 7), and the experimental procedure such as the cooling rate, the number of crystals in the solution, and the size and shape of the crystals. Various additives which may affect protein shape and stability can also lead to changes in TH [31–33]. It is important to keep all the parameters of the experiment as constant as possible. Notably, salts at low and moderate concentrations usually have minor effect on the TH but will affect the T_m and T_f in a colligative manner (i.e., shift the TH gap). High (typically a few molar) salt concentrations will lower the T_m and T_f to different extent, such that the difference between them will increase as a function of salt concentration [34].

8. **Cooling rate and time**. The time that passes from the incubation period to the end of the cooling should be considered as accumulation time. To avoid long cooling periods, we recommend a decreasing rate of – 0.1 °C per 4 seconds or – 0.5 °C/min for concentrated samples of hyperactive IBPs, with expected TH of >3 °C. For IBPs with low TH activity or dilute samples, we recommend slower ramp of 0.01 °C per 10 seconds. If the range of activities cannot be estimated, use a ramp of –0.1 °C/20 s.

9. **The effect of crystal size on the melting point**. The Gibbs–Thomson equation relates the curvature of a crystal surface to the equilibrium melting temperature. Large curvatures lead to lowering of the melting point, and therefore, the smaller the crystal, the lower the T_m. For example, the melting point of a crystal with a radius of 50 μm is –0.001 °C, while a crystal with a radius of 10 μm has T_m of –0.005 °C [35]. Thus, shrinking of the crystal reduces the melting point in a positive feedback loop. The melting temperature of an ice crystal of typically 30 μm in pure water that is used for calibration is close to 0 °C, but due to the finite size of the crystal, it is estimated to be –0.002 °C.

10. **Thermistor resistance**. The measured resistance of the thermistor at melting point of the ice is dependent on the current that is used in the measurement. A typical measurement current is 10 or 100 μA, and for a typical resistance of 30 kΩ at melting (0 °C), the power that heats thermistor, I^2R, is 3 μW

and 300 μW, respectively. The higher heat generation of the thermistor at the higher measurement current results in difference in the resistance in the calibration at melting point.

11. **The Steinhart–Hart relation for a thermistor**. This resistance as a function of temperature of a thermistor depends on the properties of the thermistor embedded in the cold stage such as the semiconducting material composing it. The Steinhart–Hart relation defines a correlation between the resistivity of the thermistor and the absolute temperature using three parameters a, b, and c [36]: $T^{-1} = a + b\ log\ (R/R25) + c(logR/R25)^3$. A modified version of the equation in an inverse form of the equation with four parameters is [37] $R(T) = R(25\ °\ C) * Exp(\ A + B/(T + 273.15) + C/(T + 273.15)^2 + D/(T + 273.15)^3)$. These four parameters compose the permanent set point of the calibration of our osmometer and are set by the manufacturer of the thermistor. They need re-setting in the LabVIEW interface only when the thermistor of the system is replaced.
12. **The significance of temperature calibration for TH and osmolality measurements**. Temperature calibration does not affect the TH since it is a relative measurement, but it is important to set an accurate 0 °C for osmolality measurements. The 0 °C should be calibrated occasionally or when the melting point is shifted more than expected (*see* **Note** 7). For accuracy, it is advisable to use pure water, double-distilled with resistance as close to 18.2 MΩ as possible. It is possible to use standard salt solutions and calibrate the osmometer at a different temperature. However, since measurements are usually conducted around 0 °C, this temperature is optimal for our needs.

Acknowledgments

We thank the Israel Science Foundation for financial support.

Conflict of Interest The corresponding author, Ido Braslavsky, was a co-founder of *μIce*, which produces nanoliter osmometers, with the editor of this Methods book, Ran Drori.

References

1. Scotter AJ, Marshall CB, Graham LA, Gilbert JA, Garnham CP, Davies PL (2006) The basis for hyperactivity of antifreeze proteins. Cryobiology. 53(2):229–239
2. Hansen TN, Baust JG (1988) Differential scanning calorimetric analysis of antifreeze protein activity in the common mealworm. *Tenebrio molitor.* Biochim Biophys Acta 957(2):217–221
3. Xiao-Lei Z, Tao-Tao C, Bao-Huai W, Zhi-Fen L, Yun-Biao F, Ling-Bo W et al (2001) DSC study on the thermal hysteresis activity of plant antifreeze proteins. Acta Physico-Chimica Sinica 17(01):66–69

4. Ding X, Zhang H, Liu W, Wang L, Qian H, Qi X (2014) Extraction of carrot (Daucus carota) antifreeze proteins and evaluation of their effects on frozen white salted noodles. Food Bioprocess Technol 7(3):842–852
5. Gaede-Koehler A, Kreider A, Canfield P, Kleemeier M, Grunwald I (2012) Direct measurement of the thermal hysteresis of antifreeze proteins (AFPs) using sonocrystallization. Anal Chem 84(23):10229–10235
6. Olijve LLC, Meister K, DeVries AL, Duman JG, Guo S, Bakker HJ et al (2016) Blocking rapid ice crystal growth through nonbasal plane adsorption of antifreeze proteins. Proc Natl Acad Sci USA 113(14):3740–3745
7. Braslavsky I, Drori R (2013) LabVIEW-operated novel nanoliter osmometer for ice binding protein investigations. J Vis Exp 72: e4189
8. Frick JH, Sauer JR (1973) Examination of a biological cryostat/nanoliter osmometer for use in determining the freezing point of insect hemolymph 1. Ann Entomol Soc Am 66(4): 781–783
9. Prager DJ, Bowman RL (1963) Freezing-point depression: new method for measuring ultramicro quantities of fluids. Science 142(3589): 237–239
10. 01.0001.001 - Clifton Technical Physics Biological Cryostat/Nanoliter Osmometer | Office of History, National Institutes of Health [Internet]. [cited 2022 May 11]. Available from: https://onih.pastperfectonline.com/webobject/C25D0E41-14ED-4B6A-B752-205433281972
11. Bar-Dolev M, Celik Y, Wettlaufer JS, Davies PL, Braslavsky I (2012) New insights into ice growth and melting modifications by antifreeze proteins. J R Soc Interface 9(77):3249–3259
12. Drori R, Celik Y, Davies PL, Braslavsky I (2014) Ice-binding proteins that accumulate on different ice crystal planes produce distinct thermal hysteresis dynamics. J R Soc Interface 11(98):20140526
13. Celik Y, Drori R, Pertaya-Braun N, Altan A, Barton T, Bar-Dolev M et al (2013) Microfluidic experiments reveal that antifreeze proteins bound to ice crystals suffice to prevent their growth. Proc Natl Acad Sci USA 110(4): 1309–1314
14. Bar M, Celik Y, Fass D, Braslavsky I (2008) Interactions of β-helical antifreeze protein mutants with ice. Cryst Growth Des 8(8): 2954–2963
15. Mizrahy O, Bar-Dolev M, Guy S, Braslavsky I (2013) Inhibition of ice growth and recrystallization by zirconium acetate and zirconium acetate hydroxide. PLoS ONE 8(3): e59540
16. Celik Y, Graham LA, Mok Y-F, Bar M, Davies PL, Braslavsky I (2010) Superheating of ice crystals in antifreeze protein solutions. Proc Natl Acad Sci USA 107(12):5423–5428
17. Buch JL, Ramløv H (2016) An open source cryostage and software analysis method for detection of antifreeze activity. Cryobiology. 72(3):251–257
18. Bar Dolev M, Bernheim R, Guo S, Davies PL, Braslavsky I (2016) Putting life on ice: bacteria that bind to frozen water. J R Soc Interface. 13(121)
19. Ramsay JA, Brown RHJ (1955) Simplified apparatus and procedure for freezing-point determinations upon small volumes of fluid. J Sci Instrum 32(10):372–375
20. Ramløv H, DeVries AL, Wilson PW (2005) Antifreeze glycoproteins from the Antarctic fish Dissostichus mawsoni studied by differential scanning calorimetry (DSC) in combination with nanolitre osmometry. Cryo Letters 26(2):73–84
21. Gilbard JP, Farris RL, Santamaria J (1978) Osmolarity of tear microvolumes in keratoconjunctivitis sicca. Arch Ophthalmol 96(4): 677–681
22. Chakrabartty A, Hew CL (1991) The effect of enhanced alpha-helicity on the activity of a winter flounder antifreeze polypeptide. Eur J Biochem 202(3):1057–1063
23. DeLuca CI, Comley R, Davies PL (1998) Antifreeze proteins bind independently to ice. Biophys J. 74(3):1502–1508
24. Micro Ice LtD. www.u-ice.com. Accessed 12 Apr 2022
25. Vuist JE, Schutyser MAI, Boom RM (2022) Solute inclusion during progressive freeze concentration: A state diagram approach. J Food Eng 320:110928
26. Koop T, Luo B, Tsias A, Peter T (2000) Water activity as the determinant for homogeneous ice nucleation in aqueous solutions. Nature. 406(6796):611–614
27. Bissoyi A, Reicher N, Chasnitsky M, Arad S, Koop T, Rudich Y, Braslavsky I (2019) Ice Nucleation Properties of Ice-binding Proteins from Snow Fleas. Biomolecules 9. https://doi.org/10.3390/biom9100532
28. Eickhoff L, Dreischmeier K, Zipori A, Sirotinskaya V, Adar C, Reicher N, Braslavsky I, Rudich Y, Koop T (2019) Contrasting behavior of antifreeze proteins: ice growth inhibitors and ice nucleation

promoters. J Phys Chem Lett 10:966–972. https://doi.org/10.1021/acs.jpclett.8b03719

29. Qiu Y, Hudait A, Molinero V (2019) How Size and Aggregation of Ice-Binding Proteins Control Their Ice Nucleation Efficiency. J Am Chem Soc 141(18):7439–7452
30. Liu XY, Du N (2004) Zero-sized effect of nano-particles and inverse homogeneous nucleation. Principles of freezing and antifreeze. J Biol Chem 279(7):6124–6131
31. Amornwittawat N, Wang S, Banatlao J, Chung M, Velasco E, Duman JG et al (2009) Effects of polyhydroxy compounds on beetle antifreeze protein activity. Biochim Biophys Acta 1794(2):341–346
32. DeVries AL (1971) Glycoproteins as biological antifreeze agents in Antarctic fishes. Science 172(3988):1152–1155
33. Sun Y, Giubertoni G, Bakker HJ, Liu J, Wagner M, Ng DYW et al (2021) Disaccharide residues are required for native antifreeze glycoprotein activity. Biomacromolecules. 22(6):2595–2603
34. Evans RP, Hobbs RS, Goddard SV, Fletcher GL (2007) The importance of dissolved salts to the in vivo efficacy of antifreeze proteins. Comp Biochem Physiol, Part A Mol Integr Physiol. 148(3):556–561
35. Liu Z, Muldrew K, Wan RG, Elliott JAW (2003) Measurement of freezing point depression of water in glass capillaries and the associated ice front shape. Phys Rev E Stat Nonlin Soft Matter Phys. 67(6 Pt 1):061602
36. Steinhart JS, Hart SR (1968) Calibration curves for thermistors. Deep Sea Research and Oceanographic Abstracts. 15(4):497–503
37. Steinhart-Hart Thermistor Calculator [Internet] (cited 2022 May 9) Available from: https://daycounter.com/Calculators/Steinhart-Hart-Thermistor-Calculator.phtml

Chapter 6

Evaluation of Ice Recrystallization Inhibition of Ice-Binding Proteins by Monitoring Specific Ice Crystals

Anika T. Rahman, Yasushi Ohyama, Sakae Tsuda, and Hidemasa Kondo

Abstract

Ice recrystallization is a phenomenon in which large ice crystals are formed at the expense of smaller ones. The resultant large ice crystals degrade the quality of frozen foods and cryopreserved biomaterials. To minimize freeze damage by controlling the ice recrystallization process, various compounds have been developed, including biological antifreezes, synthetic peptides, glycopeptides, polymers, and small molecules. To compare their efficiency, evaluation methods of ice recrystallization inhibition are important. This chapter describes a practical protocol to quantify the inhibition efficiency by observing specific ice crystals exhibiting uniform growth.

Key words Ice recrystallization, Antifreeze protein, Ostwald ripening

1 Introduction

Ice recrystallization (IR) is a thermodynamic process in polycrystalline ice in which large ice crystals are formed at the expense of small ones. The growing rate of IR is represented as the Ostwald ripening mechanism ($r(t)^3 = r_0^3 + kt$), where $r(t)$ is a radius of ice crystal at time t, r_0 is an initial radius, and k gives the recrystallization rate. Temperature fluctuations often promote IR during cold storage, which results in the degradation of materials, including frozen foods, cryopreserved cells, and pharmaceuticals. Antifreeze protein (AFP), also known as ice-binding proteins (IBPs), and antifreeze glycoprotein (AFGP) have been identified to exhibit ice recrystallization inhibition (IRI) [1, 2]. Furthermore, several reports have been published for the synthesized compounds AF(G)P-mimic or AF(G)P-inspired compounds, which show significant IRI activities. Therefore, quantitative evaluation of IRI activity is crucial to

---

Author contributions: A.R., Y.O., and S.T. designed research. A.R., H.K., Y.O., and S.T. performed research. A. R., S.T., and H.K. wrote the paper.

Ran Drori and Corey A. Stevens (eds.), *Ice Binding Proteins: Methods and Protocols*, Methods in Molecular Biology, vol. 2730, https://doi.org/10.1007/978-1-0716-3503-2_6,

developing IRI agents based on these natural and designed compounds. To date, several evaluation methods have been proposed, which contain two processes: developing ice crystals in the sample and the subsequent quantification of IR. Ice crystals are developed using splat cooling [2], sucrose sandwich splat [3], as well as capillary assay methods [4]. In the splat cooling method, an analyte solution is dropped onto the polished aluminum block cooled at −78 °C to prepare a thin flat ice specimen by flash freezing. The ice wafer is subsequently transferred to a microscope stage maintained at −8 °C to observe the ice recrystallization process for a certain period [2]. Convenient substitutes were proposed afterward, wherein analyte samples are dissolved in a solution with a relatively high concentration of sucrose. A sample solution is sandwiched between two glass coverslips to make a thin ice sample by the rapid freezing and employed for the microscopic observation (sucrose sandwich splat assay) [3]. Knight et al. reported that when estimating the IRI of AFP, a liquid portion should coexist with ice in the system by adding colligative antifreeze agents such as salts or sugars. Sucrose is added to the sample solution for this purpose [5]. Another method utilizes 10 μL glass capillaries containing a series of diluted analyte solution to develop ice crystals (capillary assay) [4] to find the minimum concentration of AFP required for IRI (IRI endpoint). Several methods have been developed for the quantitative assay of the IR process. The early method was reported to measure the average grain size of the largest ice crystals in the ice wafer prepared by the splat cooling method [6]. Latterly, ice crystal size in the frozen sample was measured by domain recognition software [7]. Budke et al. focused on the specific ice crystals with a circularity higher than 0.7 [8]. Olijve L. L.C et al. extracted circular objects by using an in-house program utilizing a circle Hough transform [9].

The ice crystals change their shapes in various manners with time in the recrystallization process, as shown in Fig. 1, a representative sucrose sandwich assay: (i) uniformly growing with keeping a shape similar to the original crystal, (ii) melting to shrink and finally disappeared, or (iii) merging with another ice crystal to become larger ice grains (accretive recrystallization), which tends to form an elongated ice crystals like a bottle gourd. The modes (i) and (ii) are considered migratory recrystallization [10, 11]. To simplify IRI analysis utilizing Ostwald ripening formula $r(t)^3 = r_0^3 + kt$, the ice growth is considered driven by the diffusion of water molecules between ice crystals. Therefore, only growing ice particles (i) by the water diffusion should be taken into account. Furthermore, a single snapshot of ice samples usually does not provide the mode of recrystallization whether the ice crystal is growing, shrinking, or merging. Hence, selecting specific ice grains is highly important to focus on uniformly growing ice crystals.

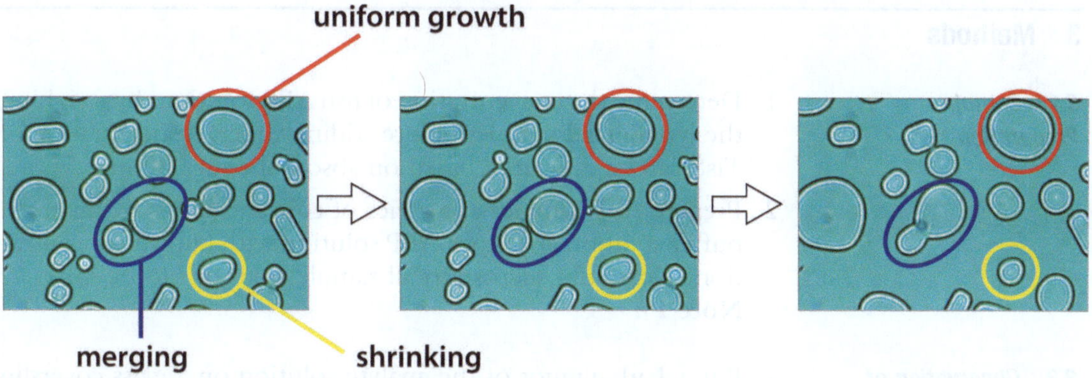

Fig. 1 Typical examples of the ice growing manner in the recrystallization process. An ice grain with uniform growth is enclosed by a red circle. Shrinking and merging ice crystals are indicated by yellow and blue circles, respectively

Recently, we reported the evaluation of IRI activity of various AFPs by measuring ice growth rate [12]. In this report, we focused on 13–15 particles of ice crystals that are growing only in mode (i). The following section describes the detailed procedure of the evaluation. In the following procedure, we use several AFPs, including fish AFP I, II, and III and microbial AFP, which exhibit different thermal hysteresis and ice-plane specificities. Fish AFP I, II, and III were prepared from fish muscle homogenate of the barfin plaice [13], longsnout poacher [14], and notched-fin eelpout [15], respectively. A defective mutant of AFP III (A20L) was also prepared, where one of the ice-binding residues (Ala 20) was replaced with Leu, resulting in reduced thermal hysteresis activity [16]. As microbial AFP, the authors prepared Tis8, which was one of the AFP isoforms secreted from snow mold fungus *Typhula ishikariensis*. Tis8 exhibited the highest thermal hysteresis activity (2.0 °C) in AFPs used in the experiment. Tis8 was prepared by recombinant expression system [17].

2 Materials

2.1 AFP Solutions

1. AFP I from barfin plaice [13].
2. AFP II from longsnout poacher [14].
3. AFP III from notched-fin eelpout [15].
4. Tis8 from *T. ishikariensis* [17].

2.2 Instrumental Setup

1. Leica DMLB 100 photomicroscope (Leica Microsystems, Wetzlar, Germany) [18].
2. Cooling stage controlled by a Linkam THMS600 temperature controller (Linkam Scientific Instruments, Surrey, UK) [18].

3 Methods

3.1 Sample Preparation

1. Determine the concentration of fish AFP solutions by weighing the lyophilized samples before adding water. Concentration of Tis8 can be estimated based on absorbance at 280 nm.
2. Prepare AFP samples at a series of concentrations by mixing a buffer solution and stock AFP solution with known concentration. Next, add sucrose to all sample solutions to be 40% (*see* **Note 1**).

3.2 Observation of Ice Recrystallization Process

1. Put a 1 μL aliquot of the analyte solution on a glass coverslip ($\phi = 13$ mm and 0.13–0.17 mm thickness). Next, sandwich the droplet of AFP solution with another coverslip with the same dimension.
2. Place the sample on the cooling stage.
3. Use an appropriate magnification lens that covers a wide sample range and a large enough sample. In our procedure, the video was recorded with 20× magnification of the objection lens, covering 240 μm (width) and 160 μm (height) in 720 × 480 pixels.
4. Decrease the temperature to −40 °C at a rate of −20 °C/min for the initial nucleation. Then, warm back the frozen sample to −6 °C at a rate of 10 °C/min to prepare a polycrystalline ice sample. The annealing temperature (−6 °C) was lower than approximately 4 °C of the melting temperature of 40% sucrose solution.
5. Record the IR process from this moment for 40 min. Repeat multiple trials for each condition.

3.3 Image Analysis

1. Select ice crystals grown uniformly in the IR process by examining each particle recorded in the video. To find such ice crystals easily, play the video in reverse from the end of the experimental period focusing on larger particles to trace the growing history of the ice crystals (*see* **Note 2**).
2. Repeat the selection processes for several observations if necessary. Select a total of 13–15 ice particles in one condition. The marked ice crystals were used for size measurement in the following process (*see* **Note 3**).
3. Capture still images from the video at 20, 25, 30, 35, and 40 min. These images are considered as images of 0, 5, 10, 15, and 20 min for the following analysis.
4. Convert each color image to 8-bit grayscale by using ImageJ [19] software (*see* **Note 4**).

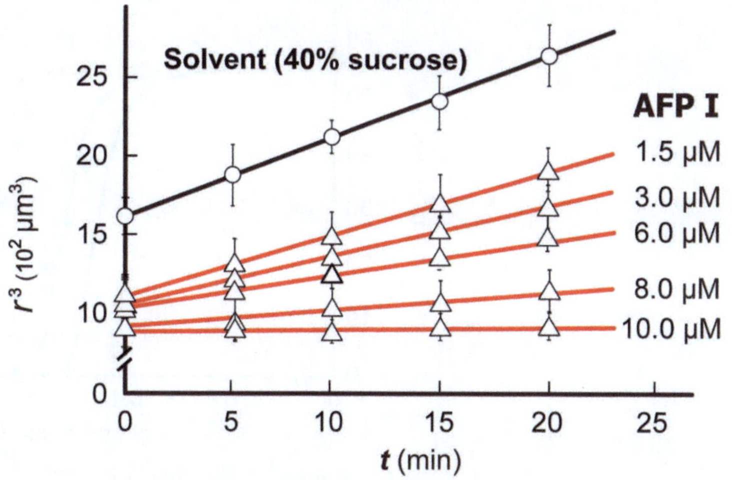

Fig. 2 An example of time dependence of the radius cubed (r^3) of a series of AFP I solutions. A solvent solution (40% sucrose) is also plotted. All of the plots show linear profiles. The slope shows the recrystallization rate k ($\mu m^3 \cdot min^{-1}$). (This figure and Fig. 3 were adopted from the published paper [10])

5. Measure the area (A) and perimeter (P) for each object in the image by using Analyze option. In this operation, apply a threshold range for the circularity of the objects to filter ice crystals in a round shape. The circularity (R) is defined as $4\pi AP^{-2}$, where $R = 1$ corresponds to a perfect circle. In our protocol, several ice crystals with R of 0.8–1.0 should be picked up.
6. Measure the particle size of ice crystals as a maximum Feret's diameter, defined in ImageJ as the longest distance between two points in the object (*see* **Note 5**).
7. Plot the cube of the mean value of r at all the subsets against annealing time at 0, 5, 10, 15, and 20 min. Figure 2 shows the representative result for AFP I solution. The plot of all AFP I concentrations shows that r^3 increased proportionately with time (t) in a linear profile, which shows that the time dependency of r^3 estimated by the protocols follows the Ostwald ripening formula, $r^3 = r_0{}^3 + kt$. The slopes of the plots are used to deduce the recrystallization rates k ($\mu m^3 \cdot min^{-1}$), which became less steep with increasing AFP concentration.
8. Estimate k values. For example, from the plot in Fig. 1, k values are estimated as 40.0, 35.0, 22.7, 9.5, and 0.0 $\mu m^3 \cdot min^{-1}$ for 1.5, 3.0, 6.0, 8.0, and 10.0 μM AFP I, respectively. Also, k for 40% sucrose solution is estimated as 57.0 $\mu m^3 \cdot min^{-1}$, corresponding to the recrystallization rate in the absence of AFP.

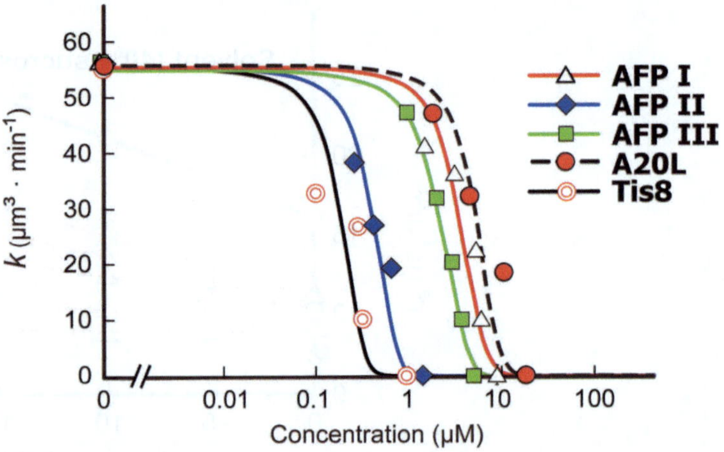

Fig. 3 The sigmoidal curves fitted into k values (μm^3·min^{-1}·M^{-1}) for the five AFP samples. The inflection point (c_i) of the curve is a parameter to compare the IRI efficiency between the samples. The IRI endpoint is estimated as a concentration at which k approaches to 0

3.4 Evaluation of Ice Recrystallization Efficiency

1. Estimate overall recrystallization rate constant $k_{I0}(c)$. Budke et al. proposed that $k_{I0}(c)$ is expressed as

$$k_{I0}(c) = k_{I0}(0) - \frac{k_{I0}(0)}{1 + \exp\left(\frac{c_i - c}{s}\right)}$$

where $k_{I0}(0)$ is the rate constant in the absence of AFP ($c = 0$). A semi-log plot version of the equation gives a sigmoidal curve with an inflection point at $c = c_i$, which is assumed to occur a turnover from diffusion-limited growth to liquid-to-ice-transfer-limited growth induced by ice binding of AFP. The parameter s defines the slope of the curve in the turnover region [7].

2. Fit the equation into plots of the observed k values (*see* **Note 6**).

4 Notes

1. IRI samples should be prepared at least 1 day before IR experiment and stored at 4 °C so that the AFP sample is fully dissolved AFP into the solvent.
2. Here, ice crystals recorded in the first half period of 20 min tended to join together with other ice, which made it difficult to find crystals without merging. In contrast, many ice crystals uniformly grew (in mode (i)) during the second half of the process, probably due to sparse and few ice crystals in the sample area. Based on these observations, pick up ice crystals from the video recorded in 20–40 min of the experimental period.

3. The 13–15 ice grains selected from each experiment subset should maintain a constant ice volume fraction with time. That will ensure that the selected ice grains are a statistical sample of the whole system. Note that the ice reshaping, which is a growth or melting of ice not due to recrystallization, often affects the ice volume fractions.
4. The threshold value for black and white should be adjusted to highlight the outline of ice crystals. Also, set a scale value between the image and the actual dimension based on values of pixel numbers and the known distance, for example, the micro ruler.
5. The value of Feret's diameter approximates the actual diameter of the round-shaped ice crystals with high circularity. To estimate the volume of each ice crystal, the shape of the crystal was approximated as a sphere with a radius (r) of half value of Feret's diameter.
6. We adjusted values of c_i and s manually by simply using spreadsheet software. The resultant plots are shown in Fig. 3, exhibiting a good agreement with observation values. Although c_i and s were not applied to optimization calculations such as least square methods, the plots show the practical trend for the IRI efficiency of various AFPs. The estimated c_i values were significantly different between the AFP samples, which were 0.27 μM (Tis8), 0.60 μM (AFP II), 3.00 μM (AFP III), 4.69 μM (AFP I), and 7.69 μM (A20L), respectively. The IRI endpoint is an approximate concentration for $k \rightarrow 0$, corresponding to the minimum AFP concentration required to halt IR. By observing the lines, the approximate endpoints were estimated as 0.7, 11.0, 1.1, 8.6, and 12.0 μM for Tis8, AFP I–III, and A20L, respectively.

Acknowledgments

This work was supported by the Japan Society for the Promotion of Science (JSPS) KAKENHI grant numbers 19H02529 and 19K22989 (for S.T.).

References

1. Capicciotti CJ, Doshi M, Ben RN (2013) Ice recrystallization inhibitors: from biological antifreezes to small molecules. In: Recent developments in the study of recrystallization. INTECH Open, Ltd., London
2. Knight CA, Hallett J, DeVries AL (1988) Solute effects on ice recrystallization: an assessment technique. Cryobiology 25:55–60
3. Smallwood M, Worrall D, Byass L, Elias L, Ashford D, Doucet CJ, Holt C, Telford J, Lillford P, Bowles DJ (1999) Isolation and characterization of a novel antifreeze protein from carrot (*Daucus carota*). Biochem J 340: 385–391
4. Tomczak MM, Marshall CB, Gilbert JA, Davies PL (2003) A facile method for

determining ice recrystallization inhibition by antifreeze proteins. Biochem Biophys Res Commun 311:1041–1046

5. Knight CA, Wen D, Laursen RA (1994) Non-equilibrium Antifreeze Peptides and the Recrystallization of Ice. Cryobiology 32:23–34
6. Eniade A, Purushotham M, Ben RN, Wang JB, Horwath K (2003) A serendipitous discovery of antifreeze protein-specific activity in C-linked antifreeze glycoprotein analogs. Cell Biochem Biophys 38:115–124
7. Jackman J, Noestheden M, Moffat D, Pezacki JP, Findlay S, Ben RN (2007) Assessing antifreeze activity of AFGP8 using domain recognition software. Biochem Biophys Res Commun 354:340–344
8. Budke C, Heggemann C, Koch M, Sewald N, Koop T (2009) Ice recrystallization kinetics in the presence of synthetic antifreeze glycoprotein analogues using the framework of LSW theory. J Phys Chem B 113:2865–2873
9. Olijve LLC, Vrielink ASO, Voets IK (2016) A simple and quantitative method to evaluate ice recrystallization kinetics using the circle Hough transform algorithm. Cryst Growth Des 16:4190–4195
10. Hartel RW (1998) The properties of water in foods. Blackie Academic and Professional, London
11. Hartel RW (2001) Crystallization in foods. Aspen Publisher, Gaithersburg
12. Rahman AT, Arai T, Yamauchi A, Miura A, Kondo H, Ohyama Y, Tsuda S (2019) Ice recrystallization is strongly inhibited when antifreeze proteins bind to multiple ice planes. Sci Rep 9:2212
13. Mahatabuddin S, Hanada Y, Nishimiya Y, Kondo H, Tsuda S (2017) Concentration-dependent oligomerization of an alpha-helical antifreeze polypeptide makes it hyperactive. Sci Rep 7:42501
14. Nishimiya Y, Kondo H, Takamichi M, Sugimoto H, Suzuki M, Miura A, Tsuda S (2008) Crystal structure and mutational analysis of Ca^{2+}-independent type II antifreeze protein from Longsnout poacher, *Brachyopsis rostratus*. J Mol Biol 382:734–746
15. Nishimiya Y, Sato R, Takamichi M, Miura A, Tsuda S (2005) Co-operative effect of the isoforms of type III antifreeze protein expressed in Notched-fin eelpout, *Zoarces elongatus Kner*. FEBS J 272:482–492
16. Mahatabuddin S, Fukami D, Arai T, Nishimiya Y, Shimiyu R, Shibayaki C, Kondo H, Adachi M, Tsuda S (2018) Polypentagonal ice-like water networks emerge solely in an activity-improved variant of ice-binding protein. Proc Natl Acad Sci USA 115:5456–5461
17. Cheng J, Hanada Y, Miura A, Tsuda S, Kondo H (2016) Hydrophobic ice-binding sites confer hyperactivity of an antifreeze protein from a snow mold fungus. Biochem J 473:4011–4026
18. Takamichi M, Nishimiya Y, Miura A, Tsuda S (2007) Effect of annealing time of an ice crystal on the activity of type III antifreeze protein. FEBS J 274:6469–6476
19. Rasband WS (1997–2018) ImageJ. U.-S. National Institutes of Health, Bethesda. https://imagej.nih.gov/ij/

Chapter 7

Measurement of Ice Nucleation Activity of Biological Samples

Rosemary J. Eufemio, Ralph Schwidetzky, and Konrad Meister

Abstract

Experimentation with ice-nucleating biomolecules is needed to advance the fundamental understanding of biotic heterogeneous ice nucleation. Standard experimental procedures vary with sample type. Here we describe a generalized primary purification and analysis process to measure ice nucleation activity of biological samples using an advanced freezing droplet assay.

Key words Heterogeneous ice nucleation, Ice-nucleating biomolecules

1 Introduction

Ice formation is the most prevalent liquid-to-solid phase transition on earth and is crucial for fields as diverse as cryobiology, geology, and climate science. At ambient conditions, the formation of ice from water is thermodynamically favored at temperatures below 0 °C, but this crystallization process is kinetically hindered. Hence, pure water can be supercooled to temperatures as low as −38 °C, below which homogenous ice nucleation occurs. In natural systems, water freezes predominately in a heterogeneous process, facilitated by the presence of ice-nucleating agents of biological and abiotic origins [1]. Naturally occurring abiotic ice nucleators (e.g., dust, minerals, clay) typically elevate freezing temperatures to −15 to −30 °C, whereas biological ice nucleators are more active and facilitate freezing at temperatures between 0 and −15 °C [2].

Nature provides us with extraordinary examples of how to induce ice formation with high efficiency. Organisms inhabiting cold environments contain biological ice nucleators (INs) that facilitate ice formation at temperatures close to 0 °C [3, 4]. The success of biological INs as an efficient protection against

Ran Drori and Corey A. Stevens (eds.), *Ice Binding Proteins: Methods and Protocols*, Methods in Molecular Biology, vol. 2730, https://doi.org/10.1007/978-1-0716-3503-2_7,

uncontrolled freezing can be witnessed by their wide distribution among different organisms including bacteria, fungi, plants, insects, and lichen [5].

Measuring the activity of the biological INs is key for understanding and modeling their role in biological and atmospheric cycles. Different instruments have been used for the analysis of INs in immersion freezing experiments, including cloud chambers, continuous-flow diffusion chambers, and freezing droplet assays [6–10]. Freezing droplet assays are particularly important since they are capable of measuring very small IN concentrations in environmental samples that are active at temperatures above −10 °C [7].

The operating principle of freezing droplet assays is the simultaneous cooling of a defined number of aqueous droplets with equal volumes in the pico- to milliliter range. The freezing experiments are usually performed using a stepped temperature profile or a constant cooling rate, which ideally should be similar to those in atmospheric or biological environments where precipitation or extracellular freezing is triggered by the formation of ice crystals. Droplet freezing can be detected using digital cameras based on the reduction of light transmission upon freezing [6, 7] or infrared cameras that detect the latent heat release upon the phase change from liquid water to ice [10–12]. The determination of frozen droplets at a given temperature or after a certain time interval then enables the quantitative assessment of INs, which was established by Vali in 1971 [13]. Given the high sensitivity of the freezing process to impurities, it is extremely important to highlight that problems can arise when testing biological samples unless strict control measures are applied [14]. Impurities in materials and solvents pose risks to the integrity of the purification process, so it is critical that for all measurements, especially biological samples, the external conditions must be controlled and documented. The lab workspace and materials must be sterile, and increased efforts should be made to minimize sources of contamination in solvents, as even minor changes in pH, storing temperature, or buffer conditions can drastically alter the experimental results [15–18].

2 Materials

Prepare all samples using pure water from a water purification system such as Millipore Milli-Q® Integral 3 or Barnstead™ GenPure™ xCAD Plus. Autoclave the water at 121 °C for 15 min and filter through a 0.1 μm-pore-size bottle top filtration unit prior to use [10, 19]. As even lowest concentrations of contaminants can dramatically affect ice nucleation efficiency, impurities represent the highest risk for the experimental determination of the nucleation temperature [14]. Highly active biological ice nucleators tend to stick to surfaces and can contaminate abiotic samples. To minimize this risk of contamination, it is necessary to disinfect the bench, fume hood, and any materials, such as spatulas, before starting the sample preparation. Consumables (e.g., pipette tips, well plates) should be used directly after opening and ideally have PCR quality. It is further recommended to measure water and control samples when switching to consumables from a new provider. To exclude contaminations in the sample preparation, the ice nucleation activity of water or solvents (buffer, salt solution) must be determined in parallel. To exclude impurities in the solvent, it should be autoclaved as described above. It is further recommended that commercially available buffer solutions should be used. Salts, buffers, and other chemicals used for sample preparation must be of the highest available grade and with known formulations.

2.1 Sample Purification

1. Sterile 50 mL centrifuge tubes (autoclaved at 121 °C for 15–30 min).
2. 0.22 μm sterile membrane syringe filter with appropriately sized syringes or 0.1 μm polyethersulfone membrane bottle top filtration units.
3. Sterile Eppendorf or centrifuge tubes for filtrate collection, storage, and subsequent analysis.

2.2 Freezing Droplet Experiments

1. Sterile Eppendorf tubes (autoclaved at 121 °C for 15–30 min).
2. Two 384-well multiwell plates of PCR quality.
3. Sterile pipette tips.
4. Liquid handling station (e.g., epMotion ep5073, Eppendorf, Hamburg, Germany).
5. Centrifuge (e.g., MPS 1000 Mini Plate Spinner, Axon Labortechnik GmbH, Germany).
6. Two optical light sources or infrared cameras (e.g., Seek Thermal Compact XR, Seek Thermal, Inc., Santa Barbara, CA, USA).

3 Methods

Procedures are typically carried out at room temperature and in a sterile workplace on a bench disinfected with 70% ethanol. Use a freezing droplet assay (described in Sect. 3.2 "Freezing Droplet Experiments") to test the background freezing temperature of autoclaved pure water prior to use. Materials (e.g., filters, gloves) may contain particles that act as ice-nucleating agents, so perform blank test runs with all materials to quantify and minimize background laboratory contamination [14].

3.1 Sample Purification

1. Collect a known mass of a biological sample in a sterile 50 mL tube.
2. Wash the biological sample (e.g., plant leaf, berries) with water to remove any impurities located on the sample. Keep the wash water for comparison.
3. If working with highly structured or intact biological samples, such as vegetation fragments, grind the samples in a mortar and pestle or homogenize them in a tissue grinder to break down cell walls [15].
4. Add enough pure water to achieve the desired concentration of the sample, e.g., 40 mg sample in 10 mL water [20]. Keep exact notes about weight and added volume.
5. Centrifuge or vortex the samples. Adapt the procedure to specific biological samples. Examples are:
   - Centrifuge vegetative fragments at 27000 × *g* for 10 min [15].
   - Vortex fungal mycelium three times at 2700 rpm for 1 min [21].
   - Manually shake suspended pollen grains for 5 min, and then stir with a stirring bar for approximately 60 mins [20].
6. If the supernatant is not clear after centrifuging, decant and centrifuge the supernatant again. It should be close to transparent after the second centrifugation.
7. 6. Filter the supernatant into a sterile Eppendorf or centrifuge tube, depending on the volume of filtrate collected. Effective filtration methods include 0.22 μm syringe filters or 0.1 μm bottle top filtration units. We find that the 0.22 μm filters often clog and it is necessary to use more than one per sample.
8. 7. Obtain at least 1.0 mL of filtrate, as this is the minimum volume needed for droplet freezing measurements with robust statistics [10]. The filtrate will contain ice nucleators from the sample. Do not reuse filters for different samples as biomolecules will stick in the filtrate and contaminate subsequent samples.
9. 8. If ice nucleation activity is not tested immediately after purification, store the filtrate at 2–8 °C for a short time or at −18 °C for prolonged storage to prevent aging or degradation. Repetitive thawing and freezing should be avoided unless the biological INs remain stable.

3.2 Freezing Droplet Experiments

1. Use a freezing droplet assay that enables robust statistics (e.g., the high-throughput Twin-plate Ice Nucleation Assay (TINA)) [10]. Use a liquid handling station to serially dilute the sample. The fully automated experimental procedure will dilute

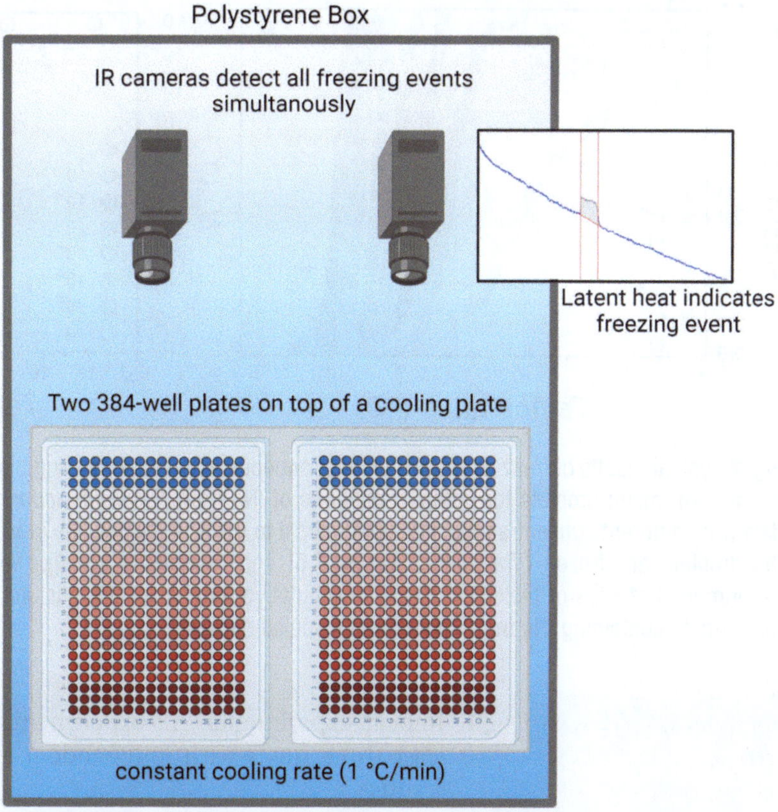

Fig. 1 Schematic illustration of a typical freezing droplet assay: two multiwell plates with droplets containing different ice nucleator concentrations, a cooling plate, and infrared cameras within a polystyrene box. IR cameras detect latent heat release during freezing

samples, with each dilution consisting of 96 droplets placed in two 384-well plates. Droplet volumes should be consistent and ideally 3 μL. Alternatively, multi-pipettors can be used, but great care must be taken that sample volumes are consistent. For initial measurements, it is advantageous to start with the highest available sample concentration. Subsequent sets of experiments can be performed with specific concentrations.

2. Briefly centrifuge the multiwell plates to ensure that droplets are equally distributed in the plates.

3. Cool the 384-well plates at a continuous rate of 1 °C/min from 0 °C to −30 °C. Use infrared cameras or optical light sources to monitor the droplet freezing and to determine the fraction of frozen droplets. A typical cooling system setup is shown in Fig. 1. For new freezing droplet assays, it is imperative to check whether the temperature gradient is similar at all positions of the multiwall plates [10].

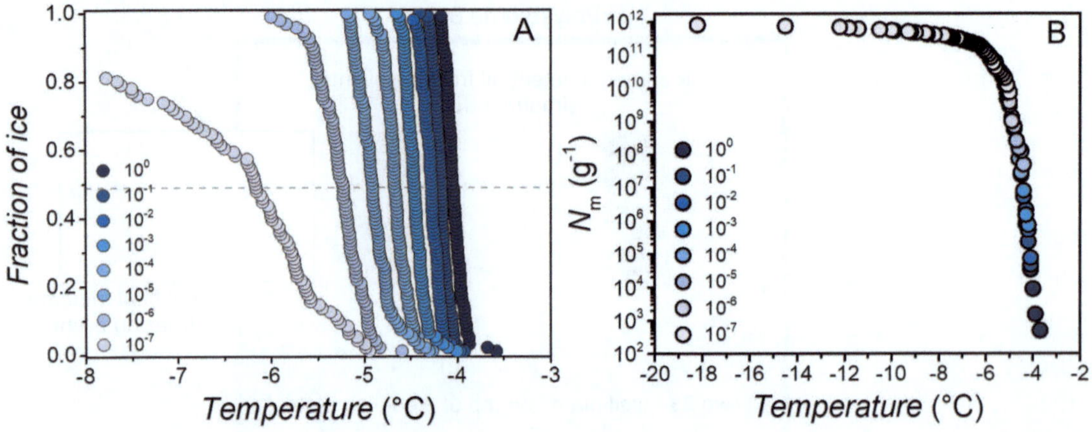

Fig. 2 Typical results of freezing experiments of aqueous extracts containing biological ice nucleators (INs). (**a**) Fraction of frozen droplets for different dilutions of INs from *Fusarium acuminatum*. Symbol colors indicate data from different concentrations and are identical to (**b**). The dashed line crosses the points at which 50% of the droplets are frozen (*fraction of ice* = 0.5) and represents the T_{50} value that is often reported in experimental studies of biological INs. (**b**) Cumulative number of INs per unit mass of *F. acuminatum* (N_m) for extracts containing INs from spores and mycelial surfaces

4. Repeat experiments. All experiments should be performed at least three times with independent samples to obtain reliable results.

3.3 Analysis of Experiments

Extract the nucleation temperatures for each droplet. Great care must be taken that the temperature gradient is equal within the cooling plate. If the temperature gradient is not similar, then temperature corrections are needed. Sort the values from the highest temperatures to the lowest temperature.

1. Calculate the fraction of ice (f_{ice}), which is dependent on the number of droplets (f_{ice} = number of frozen droplets/total number). The sorted values represent the number of frozen droplets. Plot the fraction of ice vs the temperature, as shown in Fig. 2a.
2. Determine the T_{50} value of your samples. The points at which 50% of the droplets are frozen ($f_{ice} = 0.5$) represent the T_{50} value that is often reported in experimental studies of biological INs.
3. Determine the cumulative ice nucleator number (N_m) of your samples by using the fractions of ice and Vali's equation [13] to calculate the number of active ice nucleators per mass unit of the sample. The calculation assumes that ice nucleation is a time-independent (singular) process, and a typical plot is shown in Fig. 2b.

Acknowledgments

This material is based upon the work supported by the National Science Foundation under grant no. NSF 2116528, 2308172.

References

1. Koop T, Luo B, Tsias A, Peter T (2000) Nature 406:611–614
2. Murray BJ, O'Sullivan D, Atkinson JD, Webb ME (2012) Chem Soc Rev 41:6519–6554
3. Maki LR, Galyan EL, Chang-Chien MM, Caldwell DR (1974) Appl Microbiol 28:456–459
4. Pouleur S, Richard C, Martin JG, Antoun H (1992) Appl Environ Microbiol 58:2960–2964
5. Bowles DJ, Lillford PJ, Rees DA, Shanks IA, Lundheim R (2002) Philos Trans R Soc Lond B Biol Sci 357:937–943
6. Budke C, Koop T (2015) Atmos Meas Tech 8: 689–703
7. Stopelli E, Conen F, Zimmermann L, Alewell C, Morris CE (2014) Atmos Meas Tech 7:129–134
8. Whale TF, Murray BJ, O'Sullivan D, Wilson TW, Umo NS, Baustian KJ, Atkinson JD, Workneh DA, Morris GJ (2015) Atmos Meas Tech 8:2437–2447
9. Garimella S, Kristensen TB, Ignatius K, Welti A, Voigtländer J, Kulkarni GR, Sagan F, Kok GL, Dorsey J, Nichman L, Rothenberg DA, Rösch M, Kirchgäßner ACR, Ladkin R, Wex H, Wilson TW, Ladino LA, Abbatt JPD, Stetzer O, Lohmann U, Stratmann F, Cziczo DJ (2016) Atmos Meas Tech 9:2781–2795
10. Kunert AT, Lamneck M, Helleis F, Pöschl U, Pöhlker ML, Fröhlich-Nowoisky J (2018) Atmos Meas Tech 11:6327–6337
11. Harrison AD, Whale TF, Rutledge R, Lamb S, Tarn MD, Porter GCE, Adams MP, McQuaid JB, Morris GJ, Murray BJ (2018) Atmos Meas Tech 11:5629–5641
12. Zaragotas D, Liolios NT, Anastassopoulos E (2016) Cryobiology 72:239–243
13. Vali G (1971) J Atmos Sci 28:402–409
14. Barry KR, Hill TCJ, Jentzsch C, Moffett BF, Stratmann F, DeMott PJ (2021) Atmos Res 250:105419
15. Kieft TL, Ruscetti T (1990) J Bacteriol 172: 3519–3523
16. Schwidetzky R, Lukas M, YazdanYar A, Kunert AT, Pöschl U, Domke KF, Fröhlich-Nowoisky J, Bonn M, Koop T, Nagata Y, Meister K (2021) Chemistry – A European J 27: 7402–7407
17. Lukas M, Schwidetzky R, Kunert AT, Pöschl U, Fröhlich-Nowoisky J, Bonn M, Meister K (2020) J Am Chem Soc 142:6842–6846
18. Polen M, Lawlis E, Sullivan RC (2016) J Geophys Res Atmos 121:11,666–611,678
19. Schwidetzky R, Kunert AT, Bonn M, Pöschl U, Ramløv H, DeVries AL, Fröhlich-Nowoisky J, Meister K (2020) J Phys Chem B 124:4889–4895
20. Dreischmeier K, Budke C, Wiehemeier L, Kottke T, Koop T (2017) Sci Rep 7:41890
21. Kunert AT, Pöhlker ML, Tang K, Krevert CS, Wieder C, Speth KR, Hanson LE, Morris CE, Schmale Iii DG, Pöschl U, Fröhlich-Nowoisky J (2019) Biogeosciences 16:4647–4659

Chapter 8

Investigating the Interaction Between Ice-Binding Proteins and Ice Surfaces Using Microfluidic Devices and Cold Stages

Ran Drori

Abstract

Ice-binding proteins (IBPs) protect organisms living in sub-freezing conditions by inhibiting ice growth in fish and insects, limiting ice recrystallization in plants, and assisting bacteria to adhere to ice. The mechanisms by which these proteins bind to ice and inhibit its growth have been studied both experimentally and using molecular dynamic simulations. A unique experimental technique developed to test and characterize the interactions between IBPs and ice using a combination of a microfluidic device, cold stages with millikelvin temperature resolution, fluorescence-labeled IBPs, and fluorescence microscopy is described herein. The main advantage of this technique is the ability to exchange the solution around micron-sized ice crystals and characterize their binding to and inhibition of ice.

Key words Microfluidics, Ice-binding proteins, Antifreeze proteins, Ice growth, Temperature control

1 Introduction

This chapter describes a combination of unique home-developed methods including microfluidics and high-resolution cold stages to carefully investigate the interactions between ice-binding proteins (IBPs, and more specifically antifreeze proteins, AFPs) and ice surfaces. The growth of small ice crystals in the presence of AFPs, which irreversibly bind to the ice surface [1–3], is inhibited as further ice growth is forced to occur only between adsorbed AFP molecules [4]. According to the prevalent adsorption–inhibition model [5], after AFPs bind to ice, the curvature of the ice front becomes larger, which makes the addition of water molecules to the ice thermodynamically unfavorable (the Kelvin effect). Ice growth is thus halted until sufficient supercooling is achieved to engulf the AFP molecules, depressing the non-equilibrium freezing point of the fluid. The gap between the equilibrium melting point and the

Ran Drori and Corey A. Stevens (eds.), *Ice Binding Proteins: Methods and Protocols*, Methods in Molecular Biology, vol. 2730, https://doi.org/10.1007/978-1-0716-3503-2_8,

non-equilibrium freezing point is referred to as thermal hysteresis (TH), during which the ice crystal is neither melting nor growing [6].

The dynamic process of AFP adsorption to ice is a challenging problem to investigate experimentally. Thus, the use of microfluidic devices that allow for the formation of micron-sized single ice crystals and the exchange of solutions around them is a powerful tool. Labeling the AFPs with fluorescence dyes, either GFP (green fluorescence protein) or smaller dyes such as FITC (fluorescein isothiocyanate) or TRITC (tetramethylrhodamine), will provide sensitive imaging of the ice-adsorbed AFPs. Coupling the microfluidic approach with exquisite temperature resolution (±0.001 °C) is crucial for controlling the growth and melting of micron-sized ice crystals. Using the microfluidic method described in this chapter provided compelling evidence of irreversible binding of various AFPs to ice [1, 3], and important insights into the adsorption rates of various AFPs were revealed [7]. The procedure described herein includes the fabrication of a microfluidic device, how to securely place the device on the cold stage, and how to form ice crystals inside the microfluidic device and manipulate them using a computer program.

2 Materials

2.1 Antifreeze Proteins

1. All AFPs are dissolved in buffers with which they were purified. Antifreeze glycoproteins (AFGPs) are dissolved in double-distilled water (18 MΩ-cm). All AFPs and AFGPs are stored at +4 °C or − 20 °C. The concentrations of the proteins are adjusted before the experiment, and a range of 0.1–2 mg/mL is typically used, depending on the type of the AFP or AFGP.

2.2 Microfluidic Device Fabrication

1. Polydimethylsiloxane (PDMS) (Dow Corning Sylgard 184 silicone elastomer kit).
2. Cover slips (18 X 18 mm, 0.14 mm thick).
3. Blunt-end needles (18 gauge and 20 gauge), 90-degree bent blunt needles, 18 gauge.
4. Hot plate.
5. Tygon tubing, 1/8” ID, 3/16” OD.
6. Glass syringe.
7. 1% aqueous solution of bovine serum albumin (BSA).
8. 0.22 micron filters and plastic syringes.
9. A desiccator and a vacuum pump.
10. Plasma cleaner and an oil pump.

2.3 Cold Stage and Microscopy

1. Custom-made cold stage [8] with temperature resolution of 0.001 °C.
2. Inverted microscope.
3. Infrared laser 980 nm with a dichroic mirror.
4. sCMOS camera.
5. Temperature controller (Newport 3040 or M2Lasers Ice Bloc temperature controller).
6. Sapphire disk (25.4 mm in diameter, 0.3 mm in thickness).

3 Methods

3.1 Microfluidic Device Fabrication (Without Mold Synthesis) and Pre-freezing Preparation

1. Coat the inside of a petri dish with a piece of aluminum paper and put the prepared mold inside. To learn more about mold fabrication using lithography, see these references [9, 10].
2. Mix the curing agent and elastomer using a 1:10 weight ratio, and prepare a final volume of 30–40 mL. After making the mixture, mix thoroughly until a nearly white liquid is obtained (*see* **Note 1**).
3. Pour the PDMS mixture on the mold.
4. Place the mold in a desiccator for degassing. This step will take ~30 min, depending on the vacuum flow rate (*see* Fig. 1).
5. After all the bubbles are gone, place the mold into an oven and set the temperature to 70 °C; leave the mold in the oven for ~1 h or until the PDMS is fully cured (it should be a rubber-like material when it is fully cured).
6. Using a scalpel, carefully cut around the features in the mold to create your PDMS device. Remove the cutout and place it upside down in a new petri dish. Clean the bottom side of the PDMS cutout by attaching scotch tape and peeling it off. This

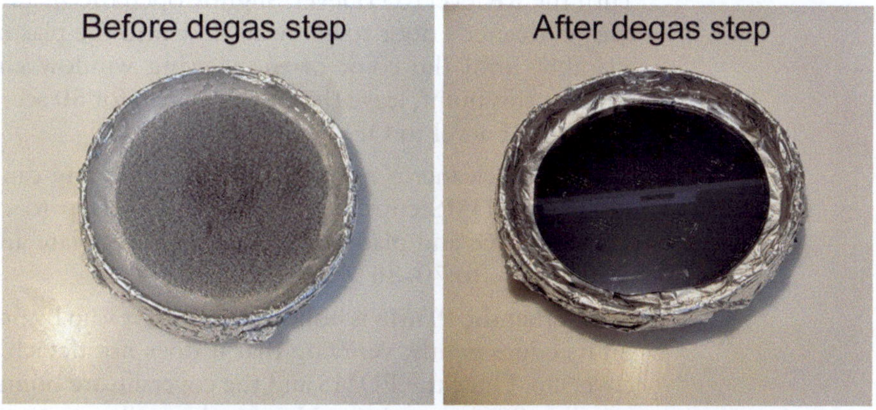

Fig. 1 Subheading 3.1, **step 4**. Degassing the PDMS that was poured on the mold (dark disk)

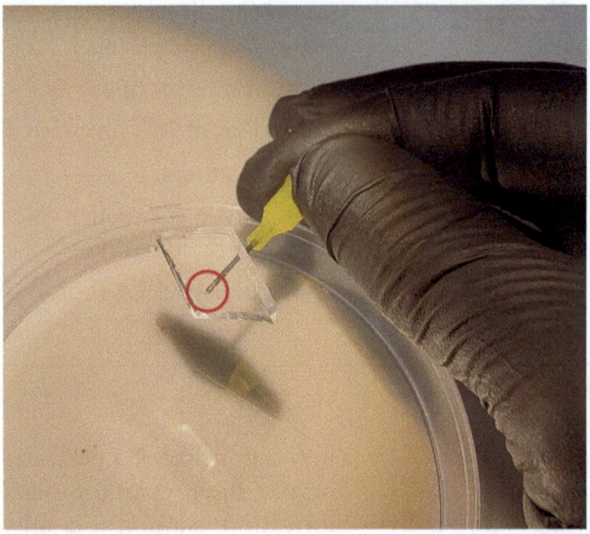

Fig. 2 Subheading 3.1, **step 7**. Punching holes for the inlets and outlets using a blunt-end needle. After a hole was punched, the PDMS piece inside the blunt needle (marked with a red circle) should be removed using tweezers

step will remove dust and other particles from the bottom surface of the cutout.

7. Using a blunt needle, punch holes in the PDMS cutout for the outlets and inlets of the channels (*see* Fig. 2). This can be done under a microscope if necessary. Remove the pieces that were punched out with tweezers.
8. Clean a coverslip (18 X 18 mm, 0.14 mm thick) using water and soap, then with isopropanol, and rinse with DI water. Clean the PDMS with scotch tape again.
9. Place the clean PDMS cutout and the clean coverslip in the plasma cleaner (*see* Fig. 3) and close the door. Power on the plasma cleaner, and start the oil pump, letting it run for 1 min. Turn the RF level to HI level. Slightly open the air valve on the plasma cleaner's door to allow air flow into the plasma cleaner chamber until the color of the viewing window changes to pink. At this point, leave the air valve open for 50 sec, and then turn the RF level to OFF.
10. If a plasma cleaner is not available, heat bonding can be used. Place the PDMS cutout on the clean coverslip to create the PDMS device and place the device on a hot plate and set the temperature to 70–80 °C, and leave for ~1 h.
11. Check that the PDMS is bonded to the coverslip by pressing up on its edges gently, verifying that it does not detach from the coverslip. Once the PDMS and the coverslip are bound tightly, use the 90-degree bent blunt-end needles to connect the

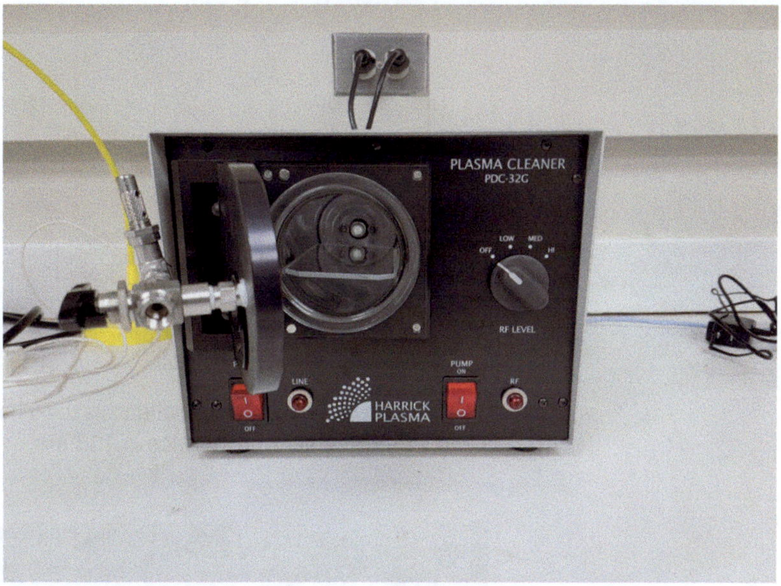

Fig. 3 Plasma cleaner

tubing to the PDMS device by removing the plastic syringe connector. This step can be done using the flat part of small nose pliers to hold the metal part of the blunt needle and pulling the plastic part of the needle to separate these two components.

12. Connect the Tygon tube to the 90-degree bent blunt-end needle and the other side of the blunt-end needle to the holes made in the PDMS device in Subheading 3.1, **step** 7. Connect one tube to the PDMS and flush with buffer to get the air out before connecting the other needle and Tygon tube to the PDMS device. At this point, the PDMS device should be bonded to the glass coverslip and have at least one inlet and one outlet tube (*see* Fig. 4).
13. The last step in the preparation of the PDMS device is to coat the inside of the microfluidic channels with a blocking agent. One way to achieve this goal is to inject an aqueous BSA solution (1%) into the inlet and leave for 20 min. Next, flush out the BSA solution by injecting the buffer solution into the device inlet.

3.2 Placement of Microfluidic Device in Cold Stage

1. Apply a small amount of immersion oil on the cold stage and clean a sapphire disk (*see* **Note 2**). Place a sapphire disk on top of the thin oil layer. Put one drop of immersion oil on the sapphire disk and place the PDMS device on that drop. At this point, a sandwich of sapphire/oil/PDMS device should be obtained.

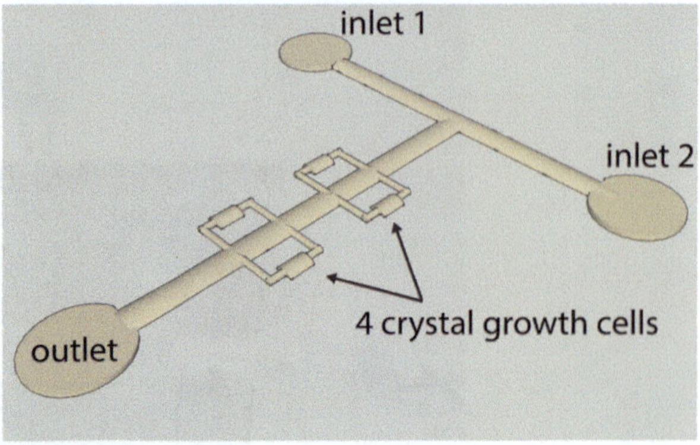

Fig. 4 Design of the microfluidic channels used here. The width of the main flow channel in the center of the device is 150 μm

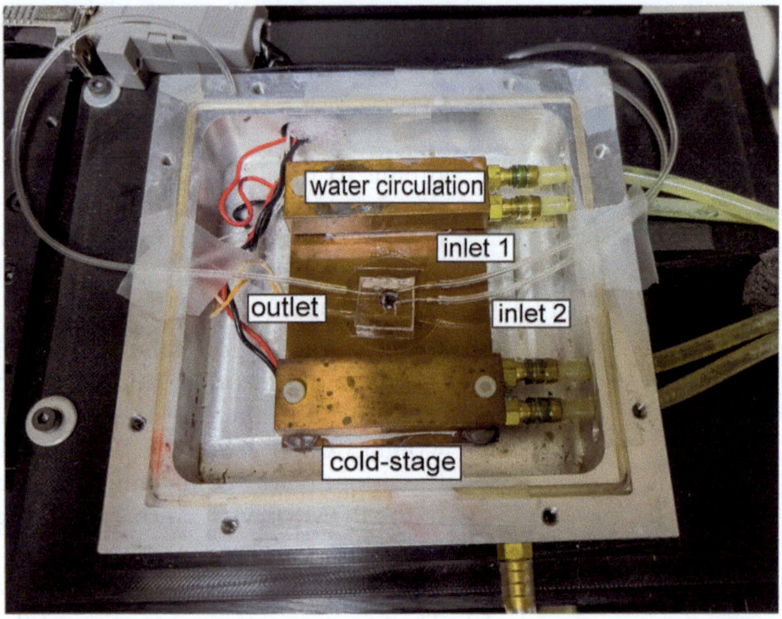

Fig. 5 The microfluidic device is placed on the cold stage, and the features (channels) are aligned with the viewing hole, allowing imaging of the ice crystals in the channels

2. Align the features (microfluidic channels) in the PDMS device with the viewing hole of the cold stage. Once aligned, hold the device in place and use scotch tape to secure the tubing of the device to the outer walls of the cold stage (*see* Fig. 5). This step will prevent movement of the device during the experiment. If the device shifts out of alignment, it can be adjusted at this point.

3. Inject the AFP/AFGP solution into the inlet using a glass syringe. A volume of 4–5 μL is sufficient. The concentration of the AFP/AFGPs can vary between 0.1 mg/mL and 2 mg/mL (*see* **Note 3**).
4. Close the lid of the cold stage.

3.3 Freezing the Sample and Obtaining a Single Ice Crystal

1. Turn on the dry air/nitrogen gas flow to keep the cold stage dry at low temperatures. Turn on the cooling bath to circulate water through the cold stage to function as a heat sink.
2. Using the temperature control program, set the sample temperature to −25 °C and observe the channels using the camera attached to the microscope. The sample will freeze at −20 to −25 °C. This freezing event will be clearly observed by the user, as the liquid in the channels will change its appearance from transparent to a rough and darker texture—ice.
3. Increase the temperature slowly (~1 °C/20 sec), slowing down the rate when reaching closer to the melting point of ice (the salts in the buffer solution will decrease the melting point by 0.5–1 °C).
4. Keep increasing the temperature until a few ice crystals remain in the desired section of the channel. Using the IR laser, melt all the ice in the inlets/outlet and in other locations in the channels.
5. After obtaining a single crystal at the desired location (*see* Fig. 6), decrease the temperature by ~0.01 °C to grow the crystal until it meets the channel walls.
6. Flow in the AFP solution using the glass syringe. The fluorescence intensity coming from the labeled AFP molecules will be dramatically increased as the AFP solution enters the microfluidic channels.

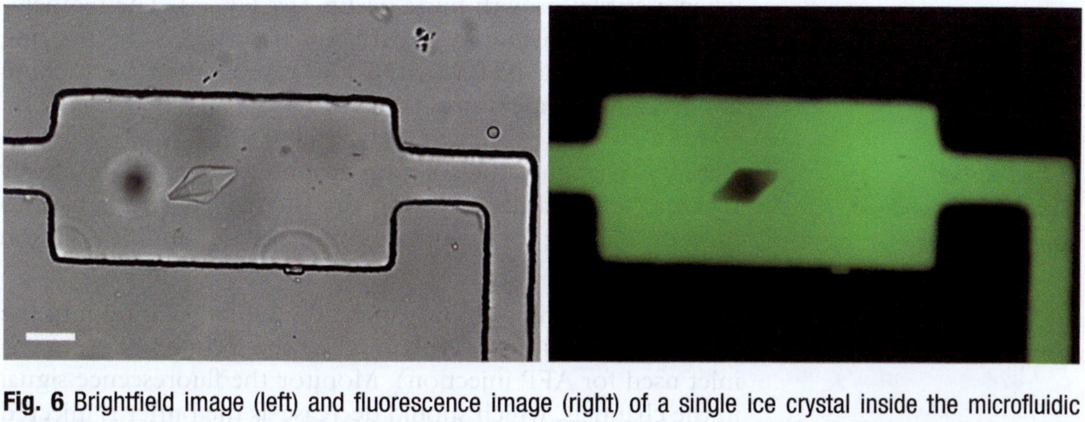

Fig. 6 Brightfield image (left) and fluorescence image (right) of a single ice crystal inside the microfluidic channels. This crystal was grown in the presence of $AFGP_{7-8}$ labeled with FITC (fluorescein isothiocyanate). Scale bar = 30 μm

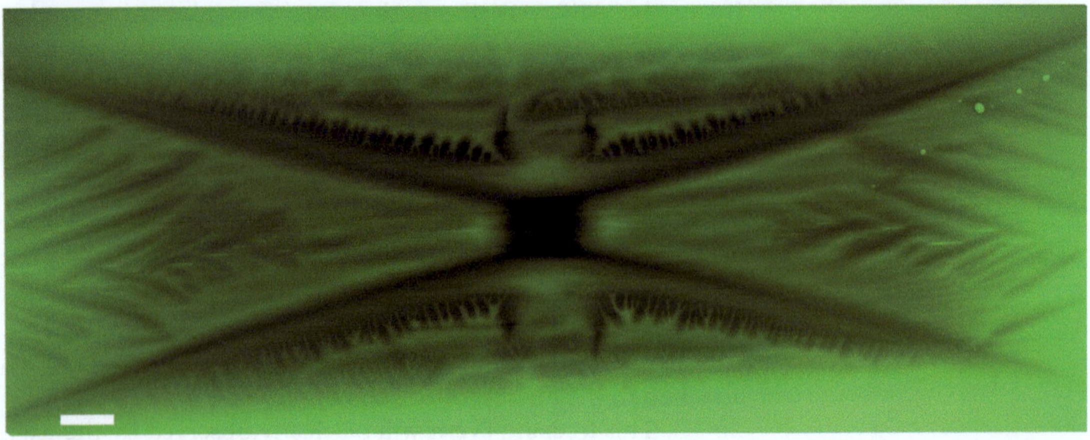

Fig. 7 Ice crystal growth burst demonstrating dendritic features. The initial crystal is located in the middle of the image, where growth started. The crystal was grown in the presence of $AFGP_{7-8}$ labeled with FITC. Scale bar = 50 μm

7. Let the AFP molecules bind the ice crystal for 5–10 min, depending on the type of the AFP. Moderate AFPs, which bind to ice fast, need about 5 min to bind to ice and saturate the surface [1, 7], while hyperactive AFPs bind slower [7], and more time (10–20 min) is needed to allow for ice adsorption (*see* **Note 4**).
8. At this point of the experiment, if needed, thermal hysteresis activity can be measured by using the RAMP function of the temperature control. First, document the melting temperature of the single crystal by slowly increasing and adjusting the temperature (small steps of 0.002 °C). The melting temperature is the highest temperature in which the crystal does not melt. Next, set the desired cooling rate (typically −0.05–0.01 °C/4 sec) and start the RAMP function. As the temperature decreases, observe the crystal and stop the RAMP function when a crystal growth burst occurs (*see* Fig. 7). The temperature at which this rapid growth was observed is the freezing temperature. The TH activity is the gap between the melting and freezing temperatures (*see* **Note 5**).

3.4 Solution Exchange around Single Crystals

1. Before exchanging the solution around a single crystal, verify that the temperature is within the thermal hysteresis gap, to prevent melting or growth of the crystal during solution exchange.
2. Using a glass syringe, slowly inject the buffer solution into the microfluidic device (*see* **Note 6**). Use the second inlet (not the inlet used for AFP injection). Monitor the fluorescence signal in the channels, which should decrease as the buffer is injected into the channels. The rate of solution exchange depends on the pressure applied on the glass syringe and should not be too fast to prevent crystal melting.

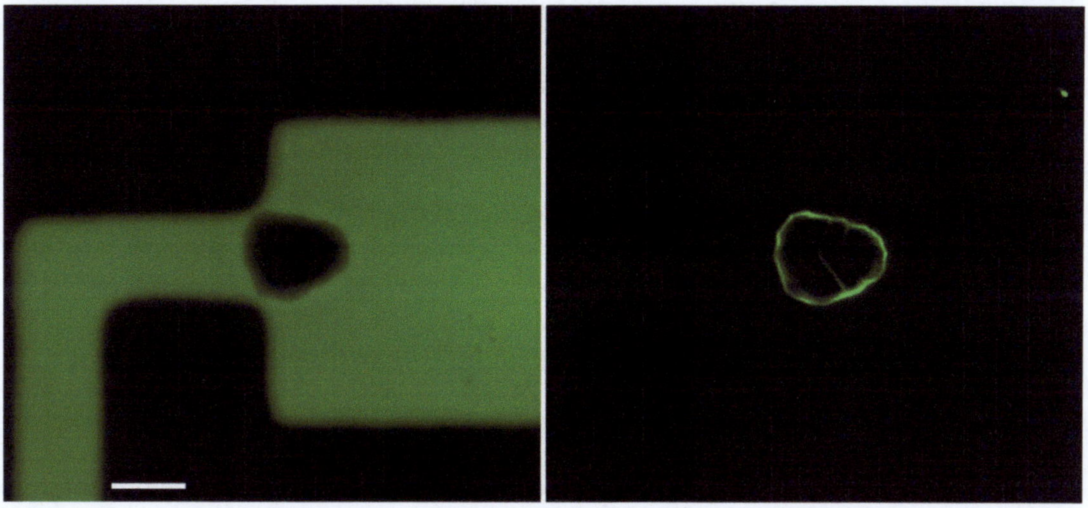

Fig. 8 Ice crystal inside the microfluidic channels before (left) and after (right) solution exchange. This crystal was grown in the presence of $AFGP_{1-5}$ labeled with FITC. The contrast of the right image was enhanced to highlight the fluorescence intensity on the ice. Scale bar = 25 μm

3. Measure the fluorescence intensity in the microfluidic channel using the imaging program. A good indication that successful solution exchange was achieved is higher fluorescence intensity on the ice surface compared to the intensity in the solution around the crystal (*see* Fig. 8).
4. To measure the TH activity after solution exchange, use the RAMP function in the temperature control program and repeat Subheading 3.3, **step 8**.
5. Obtain a new single crystal following Subheading 3.3, **step 4** and flow the AFP solution into the channels using the glass syringe. The fluorescence intensity should increase as new AFP solution flows into the channels. Repeat the solution exchange steps if needed (*see* Subheading 3.4, **steps 1** and **2**).

4 Notes

1. It is easier to pour the elastomer first and then add the curing agent, as the elastomer is very thick (similar to honey) compared to the curing agent. First, put a plastic cup on the scale and pour the elastomer into the cup. Then, add the curing agent to achieve a 1:10 weight ratio.
2. Clean the sapphire disk with soap and water and gently scrub it with your hands (disposable nitrile gloves are strongly advised). Dry the sapphire disk with a wipe or by blowing air on it. If there are still smudges, wash again, or place the disk in a beaker of soap and water and sonicate for a few minutes.

3. When injecting the AFP solution into the Tygon tubes using a syringe, add a small air bubble before and after the AFP solution to prevent dilution of this solution with the buffer. The concentrations of AFPs/AFGPs to be used are adjusted based on the fluorescence intensity, typically between 0.1 mg/mL and 2 mg/mL. Too low concentrations (> 0.05 mg/mL for most AFPs/AFGPs) might not give a strong enough fluorescence signal, and at concentrations that are too high (<2–3 mg/mL), the fluorescence signal might be quenched.
4. When deciding how long to keep the AFP solution after forming the single crystal and before exchanging the solution, the most important thing is consistency. Moderate AFPs are less sensitive to the exposure time of the crystal to the AFP solution compared to hyperactive AFPs; however, for both types, it is important to set a constant exposure time before solution exchange. Keeping a consistent exposure time will allow a more accurate comparison between the solution exchange experiments.
5. The orientation of the ice crystals in the microfluidic channels cannot be controlled, which is a limitation of this method. However, it is possible to slightly melt the crystal so that it will be free to rotate and then to lower the temperature and regrow the crystal. This solution sometimes helps in changing the orientation of the crystals in the channels. In addition, TH measurements inside the microfluidic channels are slightly limited, since sometimes the crystals are stuck between the walls of the channels and growth is mechanically inhibited.
6. Preventing the single crystal from growing or melting during the solution exchange is done by setting the temperature ~ 0.05–0.1 °C below the melting temperature of the crystal. While injecting the buffer solution into the channels, closely monitor the crystal and change the temperature slightly if the crystal grows/melts.

References

1. Meister K, DeVries AL, Bakker HJ et al (2018) Antifreeze glycoproteins bind irreversibly to ice. J Am Chem Soc 140:9365–9368
2. Knight CA, Cheng CC, DeVries AL (1991) Adsorption of alpha-helical antifreeze peptides on specific ice crystal surface planes. Biophys J 59:409–418
3. Celik Y, Drori R, Pertaya-Braun N et al (2013) Microfluidic experiments reveal that antifreeze proteins bound to ice crystals suffice to prevent their growth. Proc Natl Acad Sci U S A 110: 1309–1314
4. Knight CA, Devries AL (1994) Effects of a polymeric, nonequilibrium antifreeze upon ice growth from water. J Cryst Growth 143:301–310
5. Raymond JA, DeVries AL (1977) Adsorption inhibition as a mechanism of freezing resistance in polar fishes. Proc Natl Acad Sci U S A 74: 2589–2593
6. Scotter AJ, Marshall CB, Graham LA et al (2006) The basis for hyperactivity of antifreeze proteins. Cryobiology 53:229–239

7. Drori R, Celik Y, Davies PL et al (2014) Ice-binding proteins that accumulate on different ice crystal planes produce distinct thermal hysteresis dynamics. J R Soc Interface 11: 20140526
8. Drori R, Holmes-Cerfon M, Kahr B et al (2017) Dynamics and unsteady morphologies at ice interfaces driven by D2O-H2O exchange. Proc Natl Acad Sci U S A 114: 11627–11632
9. Qin D, Xia Y, Whitesides GM (2010) Soft lithography for micro- and nanoscale patterning. Nat Protoc 5:491–502
10. Cole RH, Tran TM, and Abate AR (2015) Double emulsion generation using a polydimethylsiloxane (PDMS) co-axial flow focus device. J Vis Exp 106:53516

Chapter 9

Quantification of the Ice Nucleation Activity of Ice-Binding Proteins Using a Microliter Droplet Freezing Experiment

Thomas F. Whale

Abstract

Many ice-binding proteins can stimulate nucleation of ice from supercooled liquid water. Quantification of ice nucleation usually involves cooling a population of droplets and observing their freezing temperatures. Here, a method for measuring the freezing temperatures of microliter-scale droplets is described and its application to quantifying a substances' ice nucleation activity discussed.

Key words Ice nucleation, Droplet freezing

1 Introduction

Contrary to common understanding, pure liquid water does not freeze at 0 °C on cooling. Observation of this phenomenon, known as undercooling or, more usually, supercooling, dates back almost as far as reliable measurement of temperature, having first been noted by Fahrenheit in 1724 [1]. For ice to form in supercooled water, a nucleation event must take place, where a cluster of ice-like molecules large enough to grow spontaneously forms. Despite the evident thermodynamic stability of ice at temperatures below 0 °C in truly pure water, this process is known to be remarkably slow, due to the large energy barrier to formation of small ice clusters. It has, for instance, been determined that a volume of water equivalent in size to Earth's hydrosphere could be held at a temperature of −20 °C for many billions of years without freezing [2], such is the improbability of the so-called homogeneous ice nucleation from supercooled water. Clearly, ice forms in the natural nvironment at temperatures above −20 °C. This is due to heterogeneous ice nucleation, nucleation induced by foreign substances in contact with supercooled liquid water. Some proteins, termed ice-nucleating proteins, are known to be exceptionally effective ice nucleators, active at temperatures close to 0 °C.

Ran Drori and Corey A. Stevens (eds.), *Ice Binding Proteins: Methods and Protocols*, Methods in Molecular Biology, vol. 2730, https://doi.org/10.1007/978-1-0716-3503-2_9,

Many substances can nucleate ice; however, direct observation of ice nucleation is not possible, as ice critical clusters are too small and rare to characterize. Subsequently, a variety of instruments have been developed to investigate and characterize this phenomenon. The majority of ice nucleation instruments have been developed in pursuit of understanding ice nucleation in clouds, which plays an important role in weather and climate [3, 4]. Additionally, various measurements have also been made on ice-binding proteins produced by cold-tolerant organisms [5]. Quantification of the ice nucleation activity (INA) of nucleators facilitates understanding of how they might behave in environmental contexts and improvements to understanding of their mechanisms of action. Experiments aimed at studying ice nucleation from supercooled water invariably observe freezing of a population of liquid droplets. Quantification of INA in this manner is not entirely routine at present. Certainly, it is not possible to buy an instrument which quantifies the INA of a substance. As such, most instruments are built by the research groups that use them.

This chapter describes a method for measuring the freezing temperatures of microliter-scale droplets using a cold plate approach. These experiments have historically often been called droplet freezing assays but are now usually referred to as droplet freezing experiments (DFEs). It should be noted that the instrument is only one of a great many developed for quantification of ice nucleation. It is based upon that described in Whale et al. [6] but uses a Peltier-driven cold stage to control temperature rather than a Stirling engine cryocooler.

The key difference between different DFEs is the size of the droplet used. The microliter-scale droplets frozen in the instrument described below are in the middle of the volume range. Many instruments use a similar scale of droplet [7–13]. Of particular interest and relevance is a study by Polen et al. [10], which examined the merits of a wide range of variations on the microliter DFE.

Recent instruments for freezing larger of 50 microliters and more have tended to use either 96-well PCR plates or flat-bottomed 96-well plates to hold water volumes [11, 14–19]. Microfluidic instruments are now often used when smaller nanoliter to picoliter droplet volumes are desired [20–24]. A comprehensive overview of DFEs in current use was recently published by Miller et al. [25].

2 Materials

2.1 Instrument Construction

The protocol described is based on the use of the microliter DFE (MDFE) Peltier-based cold stage, constructed at the University of Warwick. A very similar protocol can be used with other cold stage instruments, in particular the microliter Nucleation by Immersed

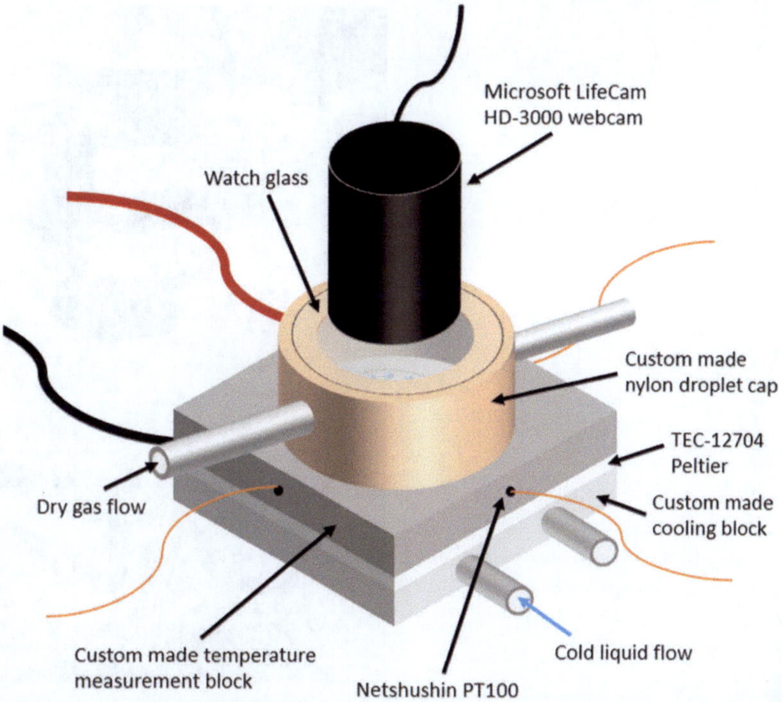

Fig. 1 Schematic of the MDFE

Particles Instrument (μL-NIPI) described in Whale et al. [6]. This section gives a brief outline of the construction of the MDFE. The core of the instrument is made up of six distinct parts, shown in Fig. 1. These are an aluminum block for cooling the hot side of the Peltier thermoelectric cooler, a TEC-12704 thermoelectric cooler, a metal plate with embedded PT100s that supports the droplets, a cap that covers the experiment and facilitates flowing dry air over it, three PT100 temperature sensors, and a camera for monitoring freezing. The arrangement of these parts is shown schematically in Fig. 1. The cooling block, temperature measurement block and top cap were all made by the University of Warwick mechanical workshop, while the remaining parts are commercially available. A retaining frame, picture in Fig. 2, is used to clamp the lid of the apparatus down, with wingnuts threading onto threaded bars to secure it. This part was also made by the University of Warwick mechanical workshop. As can be seen in Fig. 2, a retort stand and clamps are used to hold and support the instrument.

Thermal paste is used to bond the cooling block, temperature measurement block, and Peltier together (*see* **Note 1**). PT100s are coated in thermal paste and inserted into the holes in the temperature measurement block. The holes are drilled to be as close to the measurement area (the center) of the block as possible.

Fig. 2 Photograph of MDFE

One of the PT100s provides temperature measurement for a Meerstetter TEC-1091 temperature controller. This is powered by a DC power supply and connected to drive the Peltier. Peltiers are heat pumps, meaning one side heats up as the other cools when a voltage is applied across its contacts by the TEC-1091. The TEC-1091 is a proportional–integral–derivative (PID) controller and can be easily autotuned to provide temperature stability and control to the instrument (*see* **Note 2**). Heat must be removed from the hot side of the Peltier, and this is accomplished using a recirculating chiller, which passes cold fluid through the cooling block. Tap water can also be used although this limits the minimum temperature of the stage to temperatures warmer than the freezing temperature of pure water, possibly introducing difficulties. A Pico-Log PT104 Platinum Resistance Data Logger is used to read two of the three PT100s (*see* **Note 3**). Custom LabVIEW frame grabber software running on a Windows PC is used to link frames from the camera to temperatures recorded by the PT100s. As such, the instrument is capable of determining temperature of the cold stage at the time a frame is taken.

2.2 Materials for Performing a Measurement

Assuming a functional MDFE or similar DFE is available, only 22 mm silanized glass slides (Hampton Research HR3–231), pure water (*see* **Note 4** for discussion of water choice for DFEs), and an aqueous solution or suspension of the ice nucleator under investigation are needed. Pipetting is best accomplished with a Sartorius Picus Electronic micropipette or similar instrument in multi-dispense mode, although it is possible to use mechanical pipettes.

3 Methods

The method described is a general one for establishing the INA of an aqueous sample on the MDFE. Depending on the ice nucleator under investigation and the nature of its preparation (*see* **Note 5** for a discussion of sample preparation), varying combinations of measurements of backgrounds, handling blanks, and active sample may be required; however, the basic method for performing a measurement remains the same irrespective.

3.1 Performing a Measurement

1. Switch on recirculating chiller and electronic components of MDFE and set MDFE temperature to 20 °C (*see* **Note 6**).
2. Switch on instrument PC and start LabVIEW frame grabber code. It is a good idea to take a test video to ensure everything is working. On the MDFE, both measurement PT100s should read the same temperature to within 0.05 °C or so.
3. Switch on dry gas flow. For the MDFE, a flow rate of 0.2 l/min is used, although atmospheric conditions (high humidity) can require the use of higher flow rates.
4. Place 22-mm-diameter silanized glass slide (*see* **Note 7** for discussion of possible substrates) onto the center of cold stage using reversed tweezers (*see* **Note 8**).
5. Pipette droplets of pure water (for background measurements) or nucleator suspensions onto the glass slide, as shown in Fig. 3. It is best not to wear gloves as they are often packed with ice-nucleating mineral dusts that can cause contamination, although any hazards associated with sample must be considered. Typically, around 40 one μL water droplets are used, although it is possible to use both smaller and larger droplets (*see* **Note 9**). Pipetting is best accomplished with an electronic micropipette in multi-dispense mode, although it is possible to use mechanical pipettes (*see* **Note 10**).
6. Place droplet cap onto the experiment, secure with wingnuts, and set instrument temperature to 0 °C to minimize droplet evaporation while setting up experiment.
7. Start recording frames using frame grabber software, making sure to name files appropriately.
8. Manually position camera to achieve a good view of the droplet array through the instrument window. Figure 4 shows a typical array of droplets.

Fig. 3 Photograph of droplets being pipetted onto a 22 mm glass slide on the MDFE

9. Set cold stage to cool at desired rate (*see* **Note 11** for discussion of cooling rates) to a temperature below the freezing point of the droplets. Unless the droplets contain dissolved, they will certainly freeze before a temperature of −34 °C is reached. As shown in Fig. 4, the proportion of droplets frozen will increase as temperature is reduced.
10. After all droplets have frozen, set stage temperature back to 20 °C, remove top cap, and dispose of glass slide and sample.
11. The next measurement can now be performed. The MDFE often requires drying between measurement runs, either by heating to a warm temperature to speed evaporation or wiping down with lab roll.

3.2 Background Measurements

Microliter droplet freezing instruments almost invariably have a "background" meaning that droplets freeze at warmer temperatures than predicted by classical nucleation theory (CNT) for homogeneous nucleation. Trivially, if the instrument background is warmer than the nucleation induced by a sample under study,

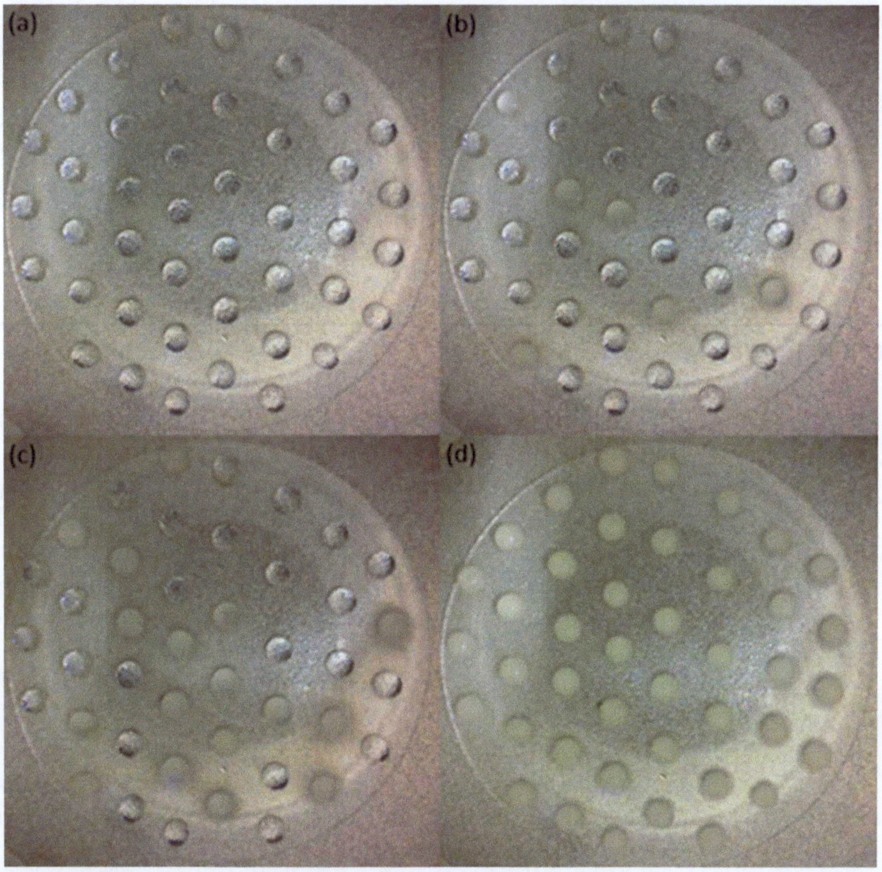

Fig. 4 Progress of droplet freezing experiment. (**a**) The start of an experiment where all droplets are unfrozen. (**b**) and (**c**) Progress of droplet freezing with decreasing temperature. (**d**) All droplets frozen at the conclusion of the experiment

then nothing can be learned about the nucleator. It is often helpful to have both a "pure water" background and a "handling" background.

3.2.1 Pure Water

A background for "pure" water on an instrument can be established by taking water from its source, be it a Milli-Q or similar ultrapure water machine, or from bottled HPLC grade water or similar and performing an experiment as described in Subheading 3.2. While there are a couple of reports of homogeneous freezing of microliter-scale droplets [8, 26], it is undoubtedly difficult to achieve [10]. Polen et al. [10] found that best results were achieved with bottled HPLC grade water filtered through a 0.22 μm filter. The issue of water purity is much more important when working with samples that nucleate at temperatures close to the instrument background. If a substance nucleates at −2 °C and the background starts at −20 °C, it is not a concern; however, if the nucleation temperatures are close to the background, then greater care must be taken.

3.2.2 Handling Blanks

As well as measurements on pure water, it is generally sensible to measure the INA of pure water subjected to whatever handling the ice nucleator undergoes, particularly if its INA is close to that of the background of the instrument. For instance, if the preparation of the sample involves a filtration step, an ice nucleation measurement should be conducted on a pure water sample passed through a similar filter, to ensure that the filtration process does not introduce INA.

4 Experiment Analysis

The MDFE produces video files with temperatures linked to each frame. For the MDFE, a custom Python code is used to facilitate visual analysis of videos. More automated methods of droplet analysis are deployed in other microliter-scale DFEs [7, 13]; however, direct analysis is common and not particularly time-consuming. The output of this process is a list of droplet freezing temperatures, which is usually presented as a droplet fraction frozen plot of the type shown in Fig. 5(a). Experiments conducted using identical conditions can be combined to increase the number of droplets available for analysis.

Analysis of ice nucleation data is a complex and debated topic. Ice nucleation is an intrinsically time-dependent or stochastic process and should in principle be described using a nucleation rate. In practice, the time dependence of ice nucleation is usually found to be minimal because individual droplets containing an identical quantity of ice nucleator generally exhibit different nucleation rates due to the existence of rare nucleation sites [27]. This means that time-independent methods of quantification of ice nucleation are commonly used. These are known as singular or deterministic descriptions [27–29]. For ice-binding proteins, the most convenient quantity to normalize to will often be mass. The cumulative number of ice nucleation active sites per unit mass of nucleator can be found using

$$n_m(T) = \frac{-\ln(1 - f(T))}{M} \tag{1}$$

where M is the mass of nucleant present in each droplet in an experiment and $f(T)$ is the droplet fraction frozen at temperature T. $n_m(T)$ can be thought of as the frequency distribution per unit mass of nucleation sites causing nucleation at temperatures higher than T. Differential spectra, which are interpreted as the number of sites causing nucleation at a given temperature rather than at all temperatures higher than T, can also be calculated but are less common in the literature [27, 30]. M can in principle be replaced by other mass-linked quantities, for instance, surface area or number of molecules per droplet, which may be more convenient or informative for some nucleators.

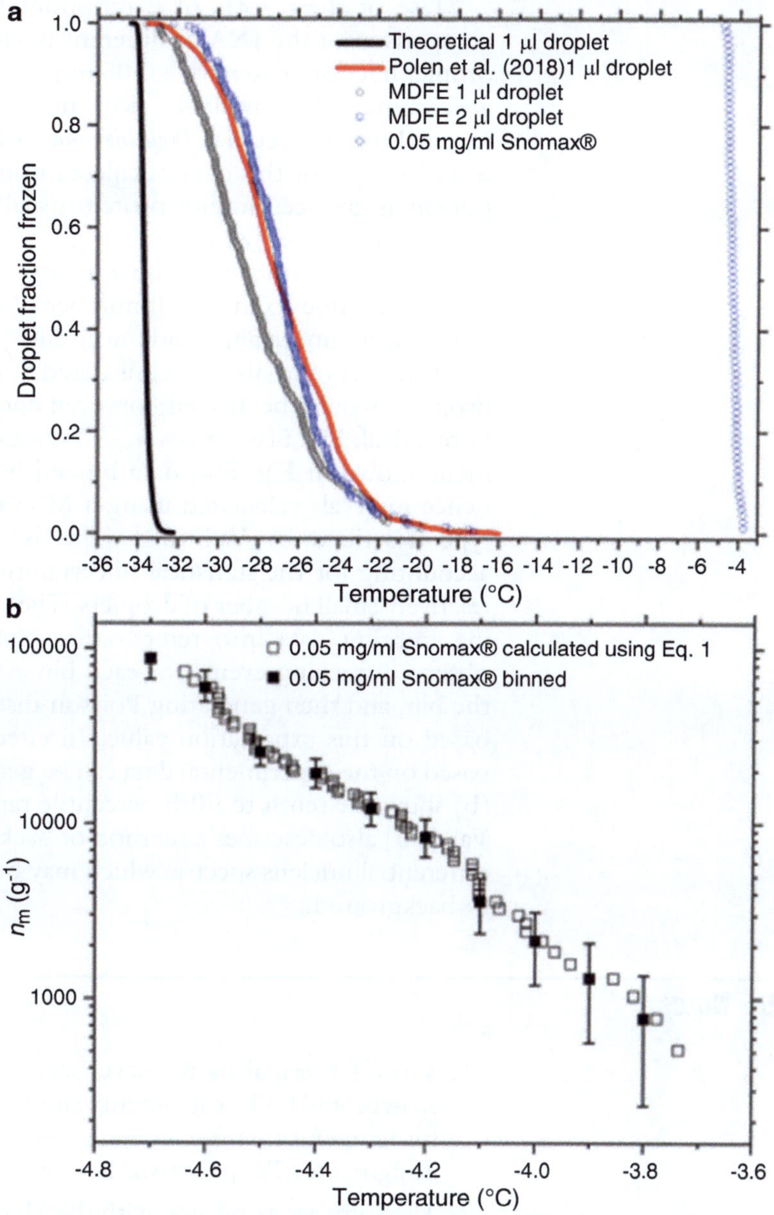

Fig. 5 (**a**) Droplet fraction frozen plots for 1 μL (273 droplets) and 2 μL (233 droplets) Milli-Q water droplets frozen on the MDFE, along with literature data for 1 μL droplets of filtered HPLC water from Polen et al. [10] and the theoretical freezing temperature of pure 1 μL given in Polen et al. [10]. Droplet fraction frozen for a 0.5 mg/mL suspension of Snomax® is also shown. (**b**) $n_m(T)$ for the Snomax® data, calculated using Eq. 1 and also binned to allow calculation of confidence intervals using a Monte-Carlo simulation, as described in the text

Use of these sorts of descriptions facilitates straightforward comparison of the INA of different nucleators. Figure 5(a) shows droplet fraction frozen for a 0.05 mg/mL suspension of Snomax®, a commercially available ice nucleator derived from the ice-nucleating bacteria *Pseudomonas syringae*. Fig. 5(b) shows $n_m(T)$ values for this data calculated using Eq. 1. The active component is the ice-binding protein usually called the bacterial ice nucleation protein [31].

In general, the main uncertainty in DFEs is the statistical uncertainty due to the small number of droplets frozen, although temperature uncertainty and uncertainty in the mass per droplet or equivalent should also be considered. Ideally, many thousands of droplets would be frozen, or even more, but this is generally impractical. Fig. 5(b) shows $n_m(T)$ values for the Snomax® experiment shown in Fig. 5(a) data binned by temperature with confidence intervals calculated using a Monte-Carlo simulation of the type described by Vali [30]. This is a convenient method of accounting for the statistical uncertainty arising from the use of a relatively small number of droplets. The simulation works by dividing freezing data into temperature bins, taking the number of observed freezing events for each bin as the expectation value for the bin, and then generating Poisson-distributed random numbers based on this expectation value. In effect, simulated experiments based on the experimental data can be generated. The bars in Fig. 5 (b) show the tenth to 90th percentile range of the simulated data. Vali [30] also describes a method of background subtraction using differential nucleus spectra, which may be of use when data is close to background.

5 Notes

1. Various thermal pastes have been used satisfactorily for constructing MDFE instruments and for coating the PT100s used for temperature measurement and control. Arctic Silver MX-5, designed for PC processor heat sinks, works well.
2. The software associated with the Meerstetter TEC-1091 allows straightforward autotuning of the PID settings. Some patience may be required to allow temperature to settle before autotuning. After autotuning, temperature control should be very rapid and responsive.
3. Use of redundant PT100s allows detection of issues with temperature gradients or miscalibration. During experiments, both PRTs should read the same temperature within 0.02 °C. Temperature uncertainty is a relatively small concern for droplet freezing assays. The quoted uncertainty of the Netshushin PRTs is ±0.15 °C. This is taken to be the temperature

uncertainty of the instrument. Temperature calibration is performed by melting water and organic solvents [32]. In general PRTs are found to be very accurately calibrated. This means that the precision with which temperature accuracy can be determined by solvent melting is necessitates no adjustment to the PRT readings.

4. It is straightforward to calculate the temperature at which water droplets of a given size should freeze homogeneously. Figure 5(a) shows the theoretical homogeneous freezing temperature of 1 μL droplets of pure water, as calculated by Polen et al. [10] from the parameterization of Koop and Murray [33]. Homogeneous ice nucleation has been occasionally achieved in microliter droplets [8, 26] but in general is not observed. Figure 5(a) shows freezing temperatures for 1 μL and 2 μL of Milli-Q water on hydrophobic glass slides measured using MDFE. Also shown is data for 1 μL droplets of filtered HPLC grade water from Polen et al. [10]. The freezing temperatures in both cases are entirely typically for experiments of this type and show that even with ultrapure water, some "background" heterogeneous nucleation takes place. This largely precludes use of microliter assays for examination of nucleation at temperatures below −25 °C or so. Polen et al. [10] concluded that bottled water provided better reproducibility than Milli-Q water. Great care must be taken with interpretation of results for nucleation occurring close to the background. For such experiments, it is recommended that handling blanks or backgrounds are performed regularly to ensure contamination levels are not changing between experimental runs. As discussed briefly above, background subtraction may be possible if freezing temperatures of a sample are close to background freezing temperatures.

5. Sample preparation will depend upon the nature of the ice-nucleating substance in use; however, a few general principles should be adhered to. Each measurement run typically requires 40 to 50 μL of liquid, so enough aqueous sample must be made up to allow sufficient repeat measurements to be performed. Often, it is convenient to make up a few mL of sample in a glass vial or centrifuge tube. If possible, the concentration of nucleator should be determined. As will be discussed later, it is usual to normalize nucleation activity to either the surface area or mass of nucleator present in each droplet. Every effort should be made to ensure that the nucleator is homogeneously dispersed throughout the water used. If some droplets contain more material than others, the results will be skewed. Most ice-binding proteins are essentially soluble in water, so this will likely be less of a concern when working with these materials. Experiments conducted with powdered, insoluble nucleators require care however.

6. At 20 °C, droplets do not evaporate quickly, and condensation does not tend to form on the instrument, making this a convenient temperature for pipetting. This temperature may have to be adjusted depending on ambient temperature and humidity.
7. Polen et al. [10] discussed the relative merits of silanized glass, metal, silicone, and polydimethylsiloxane supporting surfaces in depth. All can be used. The silanized glass slides used in the described approach are convenient. Cleaning of the sides has been found to be unnecessary, but it should be noted that they tend to attract contamination, so the boxes containing them should be kept closed wherever possible. As mentioned above, care must be taken to ensure that background measurements are taken and are reasonable.
8. Reversed tweezers greatly facilitate handling of the small slides used. Once the slide is in place, a gentle tap on the top of the slide is advised as this can sometimes force out air trapped underneath it and prevent it from moving.
9. Droplets in microliter-scale freezing assays are on the order of a few millimeters across; this is small enough that they are in thermal equilibrium with the cold stage at the cooling rates employed. Use of larger droplets reduces the number of droplets that can be placed on a single slide. Smaller droplets have a greater tendency to evaporate and are hard to pipette accurately. Droplets in the range of 0.5 μL to 5 μL work well enough.
10. Care should be taken to make sure nucleators remain dispersed during pipetting.
11. A cooling rate of 1 °C/min is very often used for DFEs. This is usually justified as being approximately the cooling rate that droplets in mixed-phase clouds typically experience. It is also a convenient cooling rate, ensuring that experiments finish in a reasonable length of time. Faster and slower rates are possible although evaporation becomes an issue at slower rates. With faster rates, at some point, the temperature of droplets will lag significantly behind the stage temperature although it is not clear what the limit is. 10 °C/min seems like a reasonable upper limit although there is not currently extensive evidence to support this. It is often sensible to cool droplets quickly to a temperature several degrees warmer than the onset temperature of droplet freezing to save time. Isothermal experiments can also be performed, although droplets in the MDFE slowly evaporate, limiting the length of such experiments. Techniques which cover droplets in oil mostly mitigate this difficult but may be at greater risk of contamination.

Acknowledgments

TFW thanks the Leverhulme Trust and the University of Warwick for an Early Career Fellowship (ECF2018-127).

References

1. Fahrenheit DG (1724) VIII. Experimenta & observationes de congelatione aquæ in vacuo factæ. Philos Trans R Soc Lond 33(382):78–84
2. Sanz E, Vega C, Espinosa JR, Caballero-Bernal R, Abascal JLF, Valeriani C (2013) Homogeneous ice nucleation at moderate Supercooling from molecular simulation. J Am Chem Soc 135(40):15008–15017
3. Murray BJ, Carslaw KS, Field PR (2021) Opinion: cloud-phase climate feedback and the importance of ice-nucleating particles. Atmos Chem Phys 21(2):665–679
4. Kanji ZA, Ladino LA, Wex H, Boose Y, Burkert-Kohn M, Cziczo DJ, Krämer M (2017) Overview of ice nucleating particles. Meteorol. Monogr 58:1.1–1.33
5. Zachariassen KE, Kristiansen E (2000) Ice nucleation and Antinucleation in nature. Cryobiology 41(4):257–279
6. Whale TF, Murray BJ, O'Sullivan D, Wilson TW, Umo NS, Baustian KJ, Atkinson JD, Workneh DA, Morris GJ (2015) A technique for quantifying heterogeneous ice nucleation in microlitre supercooled water droplets. Atmos Meas Tech 8(6):2437–2447
7. Budke C, Koop T (2015) BINARY: an optical freezing array for assessing temperature and time dependence of heterogeneous ice nucleation. Atmos Meas Tech 8(2):689–703
8. Tobo Y (2016) An improved approach for measuring immersion freezing in large droplets over a wide temperature range. Sci Rep 6: 32930
9. Polen M, Lawlis E, Sullivan RC (2016) The unstable ice nucleation properties of Snomax® bacterial particles. J Geophys Res Atmos 121(19):11,666–611,678. https://doi.org/10.1002/2016jd025251
10. Polen M, Brubaker T, Somers J, Sullivan RC (2018) Cleaning up our water: reducing interferences from nonhomogeneous freezing of "pure" water in droplet freezing assays of ice-nucleating particles. Atmos Meas Tech 11(9):5315–5334
11. Kunert AT, Lamneck M, Helleis F, Pöschl U, Pöhlker ML, Fröhlich-Nowoisky J (2018) Twin-plate ice nucleation assay (TINA) with infrared detection for high-throughput droplet freezing experiments with biological ice nuclei in laboratory and field samples. Atmos Meas Tech 11(11):6327–6337
12. Chen J, Pei X, Wang H, Chen J, Zhu Y, Tang M, Wu Z (2018) Development, characterization, and validation of a cold stage-based ice nucleation Array (PKU-INA). Atmos 9(9): 357
13. Cook F, Lord R, Sitbon G, Stephens A, Rust A, Schwarzacher W (2020) A pyroelectric thermal sensor for automated ice nucleation detection. Atmos Meas Tech 13(5):2785–2795
14. Zaragotas D, Liolios NT, Anastassopoulos E (2016) Supercooling, ice nucleation and crystal growth: a systematic study in plant samples. Cryobiology 72(3):239–243
15. Harrison AD, Whale TF, Rutledge R, Lamb S, Tarn MD, Porter GCE, Adams MP, McQuaid JB, Morris GJ, Murray BJ (2018) An instrument for quantifying heterogeneous ice nucleation in multiwell plates using infrared emissions to detect freezing. Atmos Meas Tech 11(10):5629–5641
16. Gute E, Abbatt JPD (2020) Ice nucleating behavior of different tree pollen in the immersion mode. Atmos Environ 231:117488
17. Steinke I, Hiranuma N, Funk R, Höhler K, Tüllmann N, Umo NS, Weidler PG, Möhler O, Leisner T (2020) Complex plant-derived organic aerosol as ice-nucleating particles – more than the sums of their parts? Atmos Chem Phys 20(19):11387–11397
18. Hill TCJ, Moffett BF, DeMott PJ, Georgakopoulos DG, Stump WL, Franc GD (2014) Measurement of ice nucleation-active bacteria on plants and in precipitation by quantitative PCR. Appl Environ Microbiol 80(4): 1256–1267
19. Beall CM, Stokes MD, Hill TC, DeMott PJ, DeWald JT, Prather KA (2017) Automation and heat transfer characterization of immersion mode spectroscopy for analysis of ice nucleating particles. Atmos Meas Tech 10(7): 2613–2626
20. Stan CA, Schneider GF, Shevkoplyas SS, Hashimoto M, Ibanescu M, Wiley BJ, Whitesides GM (2009) A microfluidic apparatus for the study of ice nucleation in supercooled water drops. Lab Chip 9(16):2293–2305

21. Riechers B, Wittbracht F, Hütten A, Koop T (2013) The homogeneous ice nucleation rate of water droplets produced in a microfluidic device and the role of temperature uncertainty. Phys Chem Chem Phys 15:5873–5887
22. Tarn MD, Sikora SNF, Porter GCE, O'Sullivan D, Adams M, Whale TF, Harrison AD, Vergara-Temprado J, Wilson TW, Shim J-u, Murray BJ (2018) The study of atmospheric ice-nucleating particles via microfluidically generated droplets. Microfluid Nanofluid 22(5):52
23. Tarn MD, Sikora SNF, Porter GCE, Wyld BV, Alayof M, Reicher N, Harrison AD, Rudich Y, Shim J-u, Murray BJ (2020) On-chip analysis of atmospheric ice-nucleating particles in continuous flow. Lab Chip 20(16):2889–2910
24. Reicher N, Segev L, Rudich Y (2018) The WeIzmann Supercooled droplets observation on a microarray (WISDOM) and application for ambient dust. Atmos Meas Tech 11(1): 233–248
25. Miller AJ, Brennan KP, Mignani C, Wieder J, David RO, Borduas-Dedekind N (2021) Development of the drop freezing ice nuclei counter (FINC), intercomparison of droplet freezing techniques, and use of soluble lignin as an atmospheric ice nucleation standard. Atmos Meas Tech 14(4):3131–3151
26. Fornea AP, Brooks SD, Dooley JB, Saha A (2009) Heterogeneous freezing of ice on atmospheric aerosols containing ash, soot, and soil. J Geophys Res-Atmos 114(D13):D13201
27. Vali G (2014) Interpretation of freezing nucleation experiments: singular and stochastic; sites and surfaces. Atmos Chem Phys 14(11): 5271–5294
28. Vali G (1971) Quantitative evaluation of experimental results and the heterogeneous freezing nucleation of Supercooled liquids. J Atmos Sci 28(3):402–409
29. Vali G, DeMott PJ, Möhler O, Whale TF (2015) Technical note: a proposal for ice nucleation terminology. Atmos Chem Phys 15(18): 10263–10270
30. Vali G (2019) Revisiting the differential freezing nucleus spectra derived from drop-freezing experiments: methods of calculation, applications, and confidence limits. Atmos Meas Tech 12(2):1219–1231
31. Gurian-Sherman D, Lindow SE (1993) Bacterial ice nucleation: significance and molecular basis. FASEB J 7(14):1338–1343
32. Whale TF, Murray BJ, O'Sullivan D, Umo NS, Baustian KJ, Atkinson JD, Morris GJ (2015) A technique for quantifying heterogeneous ice nucleation in microlitre supercooled water droplets. Atmos Meas Tech 8(6):2437–2447
33. Koop T, Murray BJ (2016) A physically constrained classical description of the homogeneous nucleation of ice in water. J Chem Phys 145(21):211915

Chapter 10

Measurement of Ice-Binding Protein Activity in Highly Alkaline Environments

Elizabeth A. Delesky, Aparna J. Lobo, and Wil V. Srubar III

Abstract

Ice-binding proteins (IBPs) exhibit a disruptive potential to prevent ice growth in highly alkaline solutions and related applications. IBPs, like most proteins, are sensitive to changes in their physical environment, which can alter their physical and chemical properties and ice-binding activities. Here we describe how to investigate IBP integrity for applications in alkaline environments and discuss incorporating IBPs into portland cement paste, an alkaline ceramic, to assess the reduction of ice content and mitigation of freeze-thaw damage to the brittle cementitious matrix.

Key words Alkaline materials, Cement paste, Freeze-thaw resistance, IBP applications, IBP stability

1 Introduction

Ice-binding proteins (IBPs) are well known to mitigate ice growth in physiological environments. However, ice also forms in numerous solutions and materials with elevated pH or high ionic strength. The authors have recently shown that, despite the inherent sensitivity of IBPs to non-physiological environments, IBPs show promise as a mitigation strategy to reduce ice growth in highly alkaline, non-physiological solutions [1, 2].

Proteins are well known to denature, aggregate, or degrade in non-physiological environments [3]. IBP performance is dependent on properly folded secondary and tertiary structure [3, 4]. The stability and functionality of IBPs in non-physiological environments, which can alter these structures, thus require investigation prior to implementation of IBPs in non-neutral systems.

Portland cement concrete durability is one application area that could benefit from extending IBP applications to highly alkaline environments. Portland cement concrete is a material comprised of portland cement, water, and aggregate. Portland cement and water form a paste that binds aggregates together. Some of the initial

Ran Drori and Corey A. Stevens (eds.), *Ice Binding Proteins: Methods and Protocols*, Methods in Molecular Biology, vol. 2730, https://doi.org/10.1007/978-1-0716-3503-2_10,

water remains present within an intricate multi-scale pore system [5]. The water is highly alkaline (pH ~13). When this highly ionic pore water freezes and expands, it produces internal pressures within the brittle cementitious matrix, which can lead to micro-cracking [5, 6]. Multiple and repetitive exposures to freezing and thawing, called freeze-thaw cycling, can ultimately lead to macro-scale cracking, scaling, and failure of the concrete [5, 6].

Reducing ice growth, coalescence, and the resulting damage is, thus, beneficial to extending the service life of concrete [5–7]. However, the extreme chemical composition of concrete pore solution and its high ionic strength renders it difficult to engineer new functional additives to prevent freeze-thaw damage [8, 9].

In this work, we have outlined several methodologies for determining the structural stability and ice-binding performance integrity of IBPs in highly alkaline solutions, including size-exclusion chromatography (SEC) and SDS-PAGE for IBP primary structure, circular dichroism (CD) for IBP secondary structure, and ice recrystallization inhibition (IRI) activity for IBP performance. Once screened for activity in highly alkaline solutions, IBPs that show promising behavior at pH >12 can be incorporated into cement paste. The efficacy of IBPs to mitigate ice growth within cement paste can be measured using differential scanning calorimetry (DSC) to quantify the reduction of ice crystal volume compared to an experimental control. A modified ASTM C666 standard can also be implemented to assess the ability of IBPs to mitigate damage in cement paste cylinders subjected to multiple freeze-thaw cycles.

2 Materials

This chapter assumes the reader will have purified IBPs for further investigation.

2.1 Alkaline Solutions

To determine the effect of pH on IBP stability and performance, simple alkaline solutions will suffice. As cement and concrete pore solution is composed of many chemical constituents, complex alkaline solutions could be used to determine IBP performance in more specific environments.

2.1.1 Simple Alkaline Solutions

Due to the necessity of salt for IRI activity [9], PBS should be used as the base solvent.

1. Graduated cylinder (≤ 5 mL), stir bar, stir plate.
2. pH meter with a small probe attachment.
3. Pipettes 0.2–20 μL and tips.
4. PBS.

Table 1
Potential compositions for synthetic concrete pore solution

Solution	$Ca(OH)_2$	NaOH	KOH	NaCl	$CaSO_4$	K_2SO_4	Na_2SO_4	pH
A [11]	Saturated	–	0.10 M	0.15 M	–	–	–	12.97
B [11]	Saturated	0.1 M	–	0.15 M	–	–	–	12.99
C [11]	Saturated	–	–	0.15 M	0.037 M	0.037 M	–	12.77
D [11]	Saturated	–	–	0.15 M	0.037 M	–	0.037 M	12.71
E [8]	Saturated	0.1 M	0.2 M	0.15 M	0.1 M	–	–	12.90
F [8]	Saturated	0.2 M	0.4 M	0.15 M	–	–	–	13.30
G [8]	Saturated	0.01 M	0.01 M	0.15 M	0.2 M	–	–	12.30
H [10]	Saturated	–	–	0.5 M	–	–	–	12.36
I [10]	Saturated	–	0.2 M	0.15 M	–	–	–	13.02

Additional solution compositions can be found in the referenced publications [9, 11, 12]. It should be noted that NaCl has been included in all solutions to prevent false positives when determining IRI activity

5. High concentration IBP solution (1–2 mg/mL of IBP of interest. IBPs from *Marinomonas primoryensis* and *Shewanella frigidimarina* have been investigated [1, 2]).
6. 5 M NaOH, 1 M NaOH, and 0.1 M NaOH.
7. 1 M HCl and 0.1 M HCl.

2.1.2 Complex Alkaline Solutions

The materials listed herein are all possible chemical constituents. However, depending on the chosen solution for investigation, some or all will be used. Table 1 explicates several possible synthetic concrete pore solutions that could be synthesized with a few modifications to the protocol [9, 11, 12]· In addition to those listed below, complex alkaline solutions use the same materials as simple alkaline solutions except for PBS (*see* **Note 1**).

1. $Ca(OH)_2$.
2. KOH.
3. NaCl.
4. $CaSO_4$.
5. K_2SO_4.
6. Na_2SO_4.

2.2 IBP Integrity

Since proteins often denature, aggregate, or degrade in non-physiological environments [3] a series of characterization methods should be implemented to determine IBP integrity and activity in highly alkaline solutions to screen for their potential for further investigation within cement paste (*see* **Note 2**).

2.2.1 Size-Exclusion Chromatography to Determine Degradation or Aggregation

Materials may be dependent on the specific liquid chromatography system.

1. Liquid chromatography system (size-exclusion chromatography stack) with several detectors (multi-angle light scattering (MALS), ultraviolet (UV), refractive index (RI), viscosity, etc.) (*see* **Note 3**).
2. Tosoh TSKgel G3000SWxl size-exclusion column (*see* **Note 4**).
3. Neutral pH mobile phase that has been filtered using a 0.2 μm filter (*see* **Note 5**).
4. Bovine serum albumin (BSA) 2 mg/mL standard (Pierce).
5. 500 mL of 1 mg/mL IBP loaded into each solution of interest (*see* **Notes 6** and 7).
6. 1 mL syringes and 0.2 μm syringe filters.

2.2.2 SDS-PAGE to Determine Denaturation

1. 2–20 μL pipettes and tips.
2. 15 μL of 0.5 mg/mL IBP in alkaline solution (*see* **Note 8**).
3. Trident 4× Laemmli buffer.
4. Heat block and microtubes.
5. 4–20% denaturing TGX gel from Bio-Rad (1.0 mm × 12 well) (*see* **Note 9**).
6. 1× Tris-glycine running buffer, pH 8.8 (3 g Tris base, 14.4 g glycine, 1 g SDS in 1 L).
7. Electrophoresis apparatus and voltage source.
8. 10–250 kDa protein ladder (e.g., New England Biolabs).
9. Coomassie SimplyBlue SafeStain (Invitrogen).

2.2.3 Circular Dichroism to Determine Secondary Structure

Exact materials may be dependent on the specific CD system.

1. Circular dichroism and fluorescence spectrometer.
2. Cuvettes (1 or 0.5 mm path length).
3. 20–200 μL pipette and gel-loading pipette tips.
4. 200 μL of 0.5 mg/mL IBP in alkaline solution (*see* **Note 10**).
5. Control alkaline solution.
6. DI or Milli-Q water for dilutions.

2.2.4 IRI Activity Determination

The IRI splat assay method that will be used will be dependent on the viscosity of the solution. Materials for multiple methods are listed below.

1. Dry ice.
2. Aluminum block (approx. 12 × 12 × 5 cm).
3. Styrofoam cooler/container.

Table 2
Suggested cement mix design matrix for testing freeze-thaw resistance using IBPs in cement paste with a 0.42 solution to cement ratio. The recommended values below should yield six test cylinders (16 × 32 mm)

Mix	Cement powder (g)	IBP solution (mL)	IBP concentration (mg/mL)	IBP by wt. % of cement powder
A	72	30	0.1	0.004%
B	72	30	0.3	0.013%
C	72	30	0.5	0.021%

4. 1 m PVC pipe (inner diameter ~ 4 cm).
5. Ring stand with clamp.
6. Glass slides or coverslips.
7. 2–20 μL pipettes and tips.
8. IBP pH solution.
9. Control pH solution.
10. Sucrose.
11. Microscope with controllable cold stage or osmometer stage.
12. Microscope camera.

2.3 Cement Mix Design

The materials needed to make cement paste samples are listed below. A few suggested mix designs can be found in Table 2 (*see* **Note 11**).

1. Ordinary portland cement type I/II or CEM I/II (*see* **Note 12**).
2. DI water.
3. ~100 mL disposable container and metal scoopula (*see* **Notes 13** and **14**).
4. Plastic 16 × 32 mm (diameter x height) polyethylene cylinder molds (*see* **Note 15**).
5. Timer.
6. Airtight sealable container.
7. Sodium phosphate and container (*see* **Note 16**).
8. Razor blade.

2.4 Cement Performance Characterization

Hardened cement paste can be characterized for durability via freeze-thaw performance and for ice content via differential scanning calorimetry (DSC).

2.4.1 Freeze-Thaw Performance of Cement Paste Cylinders

1. Waterproof marker.
2. Freeze-thaw chamber (*see* **Note 17**).
3. Sealable container.
4. Ice (*see* **Note 18**).
5. Camera.
6. X-ray microscope (XRM) and micro-computed tomography system (*see* **Note 19**).

2.4.2 DSC for Ice Content Determination

1. Differential scanning calorimeter (*see* **Note 20**).
2. Hermetic sealable DSC pans.
3. Die cutter with the same diameter as the DSC pan.
4. Hammer.
5. Razor blade.
6. Tweezers.
7. 20 mL scintillation vial.

3 Methods

3.1 Alkaline Solutions

The same procedure can be followed to make both simple and complex alkaline solutions. The difference is the addition of salts for complex alkaline solutions during **step 3**. pH values are determined at ambient temperature (*see* **Note 21**).

1. Determine the target solution to create (e.g., simple alkaline pH 13 solution, complex alkaline solution from Table 1) (*see* **Notes 22**, **23**, **24**, and **25**).
2. Determine the amount of IBP stock solution to incorporate for the desired concentration in a final volume of 1 mL. Add to the graduated cylinder (*see* **Note 26**).
3. If making a complex alkaline solution, weigh out salts to be incorporated. For example, to make 1 mL of Solution B from Table 1 (saturated $Ca(OH)_2$, 0.1 M NaOH, and 0.15 M NaCl), add 1.73 mg $Ca(OH)_2$ and 8.8 mg NaCl (*see* **Notes 27** and **28**).
4. Fill the graduated cylinder to 0.7 mL using PBS (simple alkaline) or in Milli-Q/DI water (complex alkaline).
5. Check the pH using the pH meter. Add 1 M NaOH until you approach the desired alkaline pH. When ~0.2 away, switch to 0.1 M NaOH (*see* **Notes 29** and **30**).
6. Fill the graduated cylinder to 1 mL using PBS (simple alkaline) or Milli-Q/DI water (complex alkaline solution).

7. Transfer to a freezer-safe tube and label. Incubate at ambient temperature for 24 h to ensure the IBP has reached an equilibrium folding state, and then freeze until further testing.
8. Create a control solution with the same constituents as the IBP solution. Transfer to a freezer-safe tube and label (*see* **Note 31**). Freeze until further testing.

3.2 IBP Integrity

Several methods are detailed below to investigate IBP integrity in alkaline solutions. The goal of these investigations is to determine large deviations in IBP structure and performance relative to neutral solution (*see* **Note 32**). Unless otherwise specified, all IBPs should be compared to control alkaline solutions with the same composition—normalizing to each respective control solution ensures investigation of the effect of the IBP and not the variance of the pH.

3.2.1 SEC

As methods may vary depending on the specific system, a general protocol has been provided.

1. Equilibrate the SEC system with running buffer (*see* **Notes 33** and **34**).
2. Prepare the BSA standard, IBP in alkaline solutions, and control alkaline solutions for injection. Centrifuge samples to remove any large particulates. Use a 1 mL syringe to collect the top ~80% of the solution to use for analysis (*see* **Note 35**). Filter 500 μL of sample through a 0.2 μm syringe filter into a SEC-MALS sample vial (*see* **Notes 6** and **36**).
3. Create a run profile for the SEC system. While system dependent, variables to consider are injection volume, flow rate, run time, and the dn/dc of sample. Typically, a few runs are required to determine the best parameters for the specific sample, but some examples of parameters are listed below.
   (a) An injection volume of 50 μL for ~50 μg IBP/injection.
   (b) A flow rate of 0.4 mL/min.
   (c) A run time of 45 min (*see* **Note 37**).
   (d) A dn/dc of 0.185 mL/g for proteins (*see* **Note 38**).
4. Calibrate the equipment with 66.4 kDa BSA in running buffer to determine elution times in the system equilibrated with running buffer.
5. Run the samples (*see* **Note 39**).
6. Analyze the data. Subtract the baseline of the control alkaline solutions from the respective IBP in alkaline solution to reduce noise and determine the effect of pH (*see* **Note 40**). Stack the spectra and compare against IBP in its preferred solution to determine deviations in primary structure. Changes in peak

elution time with deviations by ±1 min can be attributed to pH adjustors. Greater changes in peak elution time, or changes in peak shape or intensity, can be attributed to alterations of IBP primary structure. Example of SEC-MALS data for determining deviations in primary structure is shown in Fig. 1a.

3.2.2 SDS-PAGE

1. Add 5 μL of Trident 4× Laemmli buffer to 15 μL of IBP in alkaline solution in microtubes.
2. Place samples in heat block and boil for 10 min at 95 °C with the lid of heat block set at 98 °C to prevent sample evaporation.
3. While boiling sample, set up 4–20% denaturing TGX gel from Bio-Rad into the electrophoresis apparatus.
4. Fill the inside of the gel insert with 1× Tris-glycine running buffer (*see* **Note 41**).
5. Fill the outer chamber with 1× Tris-glycine running buffer.
6. Load 5–8 μL of 10–250 kDa protein ladder into the first well. Load 10–15 μL of IBP in its preferred environment next to the protein ladder. Load 10–15 μL of IBPs in alkaline solutions in subsequent wells. Make sure the ladder and IBP samples settle to the bottom of their respective wells (*see* **Notes 42** and **43**).
7. Add the lid to the electrophoresis apparatus. Ensure the cables to the voltage controller connect positive to positive and negative to negative.
8. Run at 150 V, 300 mA, and 100 W under a constant voltage. The run time will depend on the gel and the IBP but should be between 30 and 75 min. The gel is done running when the dye bands reach the bottom of the gel (*see* **Note 44**).
9. Turn off the voltage source and remove the plugs. Remove the gel, pour out the liquid, and collect the gel from between the glass plates.
10. Clean the gel by placing it in Milli-Q water in a microwave-safe container. Microwave for 1 min, and shake for 1 min. Repeat 3×, replacing the Milli-Q water each time.
11. Replace the Milli-Q water with 1× Coomassie SimplyBlue Safe-Stain. Microwave for 1 min and shake for 2 min. Clean off excess stain by repeating **step 10** (*see* **Note 45**).
12. Image the gel to estimate molecular weight compared to the protein ladder and retention of primary structure compared to IBP in neutral solution. Due to the influence of pH adjusters, the IBP bands in the gel may appear a little deformed (e.g., "smiles" or "frowns"). The main point of investigation for SDS-PAGE is that the IBP demonstrates at the same MW band as IBP in its preferred environment. Example of SDS-PAGE data for determining deviations in primary structure can be seen in Fig. 1b.

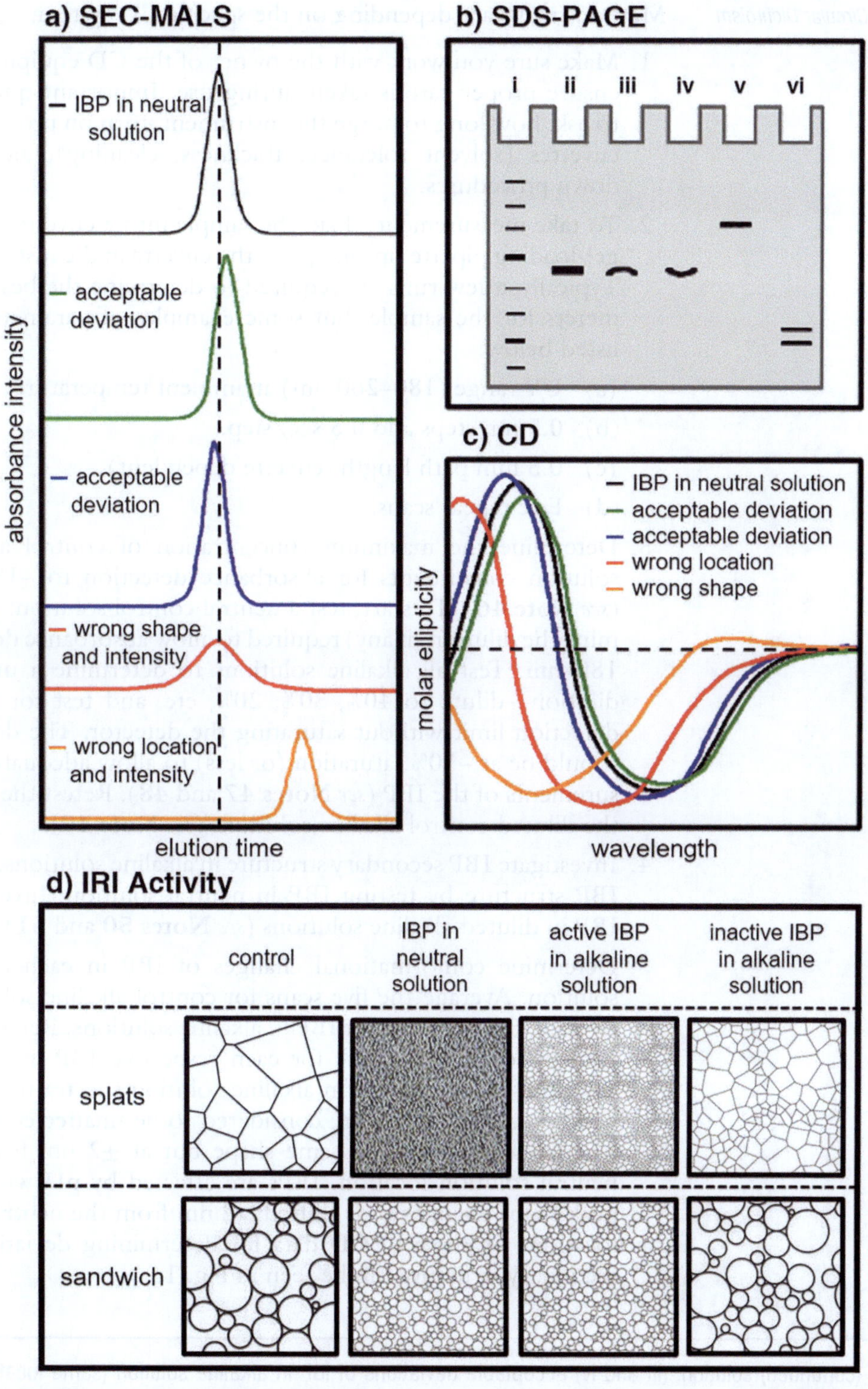

Fig. 1 Examples of data for determining changes in structural integrity for IBPs in alkaline environments compared to IBPs in neutral solution. (**a**) SEC-MALS data that demonstrates acceptable deviations (minor fluctuations in elution time, same peak shape and intensity), as well as example of deviations that indicate some significant change of structure. (**b**) SDS-PAGE data that shows (i) protein ladder, (ii) IBP in neutral

3.2.3 Circular Dichroism (CD)

Methods may vary depending on the specific CD system.

1. Make sure you work with the owner of the CD equipment to ensure proper care is taken during use. Important questions to ask: how long to purge the instrument, turn on procedures, cuvettes (solvent tolerance, thickness, cleaning), and shut down procedures.
2. To take measurements, load the sample into a cuvette using a gel-loading pipette tip, and place the cuvette in the instrument. Typically, a few runs are required to determine the best parameters for the sample, but some examples of parameters are listed below:
   (a) UV range (180–260 nm) at ambient temperature.
   (b) 0.5 nm steps and 0.5 sec/step.
   (c) 0.5 mm path length (cuvette dependent).
   (d) Five repeat scans.
3. Determine the maximum concentration of control alkaline solution constituents for absorbance detection to ~180 nm (*see* **Note 46**). To start, test a neutral control solution. Determine the dilution (if any) required to allow absorbance down to 180 nm. Test all alkaline solutions to determine a uniform dilution—dilute to 40%, 30%, 20%, etc. and test for lowest detection limit without saturating the detector. The detector should be at ~50% saturation (or less) to allow adequate measurements of the IBP (*see* **Notes 47** and **48**). Retest the pH of the diluted control alkaline solutions (*see* **Note 49**).
4. Investigate IBP secondary structure in alkaline solutions. Verify IBP structure by testing IBP in neutral solution. Investigate IBP in diluted alkaline solutions (*see* **Notes 50** and **51**).
5. Determine conformational changes of IBP in each alkaline solution. Average the five scans for control alkaline solutions. Average the five scans for IBP in alkaline solutions. Remove the control solution baseline for each respective IBP in alkaline solution. Compare IBP in alkaline solution spectra to IBP in neutral solution. IBPs are considered to be unaffected by pH when peaks exhibit the same shape but at ±2 nm from the neutral solution location. IBPs are affected by pH when the peak shape changes or is shifted >2 nm from the neutral peak location. Example of CD data for determining deviations in secondary structure can be seen in Fig. 1c.

Fig. 1 (continued) solution, (iii and iv) acceptable deviations of IBP in alkaline solution (same location, but either a "frown" or "smile"), and (v and vi) bands that indicate a lack of IBP structural integrity. (**c**) CD data that demonstrates acceptable deviations (minor deviations in intensity and peak wavelength), as well as example of deviations that indicate some degradation of structure. (**d**) IRI activity data for splats and sandwiches that provide examples of active and inactive IBPs compared to control solutions

3.2.4 IRI Activity Determination

The method for IRI activity determination will depend on the sample. For most purposes, the splat method will suffice for determining IRI activity [12]. However, samples with some precipitate or high viscosity solutions may require the sandwich method (*see* **Note 52**) [13].

1. Create serial dilutions to investigate concentration dependence and lower limit for the IRI activity of the IBP in alkaline solution (*see* **Note 53**).
2. Set up the splat apparatus. If the sandwich method is being performed, you only need to follow step 2b. Affix the PVC pipe into the clamp of the ring stand. Place at the edge of the counter so that the PVC pipe hovers over the edge of the counter and the opening of the pipe is perpendicular to the floor. The top of the PVC pipe should be between 1.5 and 2 m from the floor. Place the aluminum block inside the Styrofoam cooler with the dry ice. Place the aluminum block/cooler/dry ice on the floor underneath the PVC pipe so that the aluminum block is directly underneath the opening of the PVC pipe. Place glass slides on the aluminum block under the opening of the PVC pipe (*see* **Note 54**). Let cool for 5–15 min (*see* **Note 55**).
3. While the aluminum block is cooling, make sure the microscope and cold stage are turned on. The cold stage should reach approximately −8 °C before performing splats.
4. Create a monolayer of ice crystals using the IRI apparatus. For the splat method, dispense a 10 μL droplet of sample through the PVC pipe onto a microscope slide on the aluminum block. For the sandwich method, add sucrose to the sample to create a 30% sucrose solution (0.3 g/mL). Dispense 1 μL of sample between two coverslips and place on the aluminum block, approximately 5–10 min or until it becomes opaque.
5. Transfer the slide to a microscope cold stage to monitor recrystallization. Use the microscope to observe the monolayer of ice (*see* **Note 56**).
6. Allow the sample to recrystallize at −4 to −8 °C for 30 min (splats) or for 60–120 min (sandwich).
7. After incubation, collect images of the monolayer of ice in three different locations on the sample to observe ice recrystallization, seen as an increase in grain diameter.
8. Perform **steps 4, 5, 6,** and 7 in triplicate for each sample and control solution.
9. The images can be compared qualitatively or quantitatively. If quantitatively, count ~30 grains per splat (preferably from three different regions) by measuring the longest diameter of the grain. Fifteen of the largest grains should be counted per image and then those values sorted to take the largest ten.

Approximately 90 grains should be counted for each sample type (30/sample × 3 replicates = 90 grains). Each IBP in alkaline solution should be normalized to its respective control alkaline solution to report the IRI activity as a percentage change to understand the influence of the IBP (*see* **Note 57**) [14]· Example of IRI data for structure can be seen in Fig. 1d.

3.3 Cement Mix Design

1. Dissolve IBPs in DI water at concentrations matching the chosen mix design in Table 2.
2. Using a scale, mass 70 g of cement powder into the 100 mL disposable container.
3. Add 30 mL of IBP solution to cement powder.
4. Begin timer upon addition of IBP solution. Stir for 3 min (*see* **Note 58**).
5. Use the scoopula to fill the 16 × 32 mm plastic cylinder molds with cement paste.
6. Gently tap the molds on the benchtop so that there are minimal visible air bubbles and the top of the solution lays flat (*see* **Note 59**).
7. Place cylinder molds in the sealable plastic container.
8. Create a sodium phosphate slurry in a different small container. Fill 3/4 of the container with sodium phosphate and add tap water until a saturated solution has been made. Place the slurry in the sealable plastic container with cylinder molds (*see* **Note 60**).
9. Seal container and leave for 14 days for cement paste curing (*see* **Note 61**).
10. Remove samples from the sealed container.
11. Use the razor blade to carefully cut and peel the cylinder mold from cement paste sample.

3.4 Cement Performance Characterization

3.4.1 Freeze-Thaw Performance

1. Label all samples with waterproof marker. Ensure a control sample without IBP has been fabricated for comparison.
2. Place samples horizontally in sealable container (*see* **Note 62**).
3. Fill container with tap water so that ~1/3 of each sample (~5 mm of the diameter) is submerged.
4. Seal container and place in freeze-thaw chamber.
5. Set freeze-thaw chamber to freeze (−15 °C) for 1.5 h.
6. Set freeze-thaw chamber to thaw (16.5 °C) for 2.5 h.
7. Repeat **steps 5–6** for 30 cycles (*see* **Note 63**).
8. Remove samples from freeze-thaw chamber.

9. Visually assess external damage using a camera to qualify amounts and relative sizing of cracks. Example of schematics for comparison can be seen in Fig. 2a.
10. Internal damage can be qualified using MXCT for cross-sectional scans of the samples (*see* **Note 19**). Example of schematics for comparison can be seen in Fig. 2a.

3.4.2 DSC for Ice Content Determination

1. Using a hammer and dye cutter, break sample into 15–45 mg pieces.
2. Using the razor blade, cut down samples to fit within the DCS pan (*see* **Note 64**).
3. Saturate the cement samples by placing the cut sample pieces into a 20 mL scintillation vial filled with tap water for 2 h (*see* **Note 65**).
4. Mass the empty DSC pan and lid.
5. Remove a water-saturated sample piece and gently blot the surface dry.
6. Mass the sample piece and place into the DSC pan.
7. Hermetically seal the DSC pan and load into DSC.
8. Set DSC to equilibrate to 15 °C, cool to −60 °C, and thaw back to 15 °C with a ramp rate of 5 °C per minute.
9. Perform DSC on three replicates per sample for ice content analysis.
10. Analyze the data. Representative data for DSC curves can be seen in Fig. 2b.
    (a) A qualitative analysis of total ice content can be performed by overlapping the DSC heat flow versus temperature curves of samples and comparing the depth and width of the thawing peak.
    (b) Quantitative analysis can be performed by comparing the ice content values across different samples in the different freezing events as well as the overall ice content.
    (c) Integrate the thawing peak on the heat flow versus temperature curve and divide integrated value by heat of fusion, 332.4 J/g, to get an overall ice content.
    (d) Repeat **step 9c** for mesopore and micropore peaks to get ice content for their respective freezing events.
    (e) Subtract the mesopore and micropore ice contents from **step 11** from the overall ice content from **step 10** to get the super-cooling ice content.

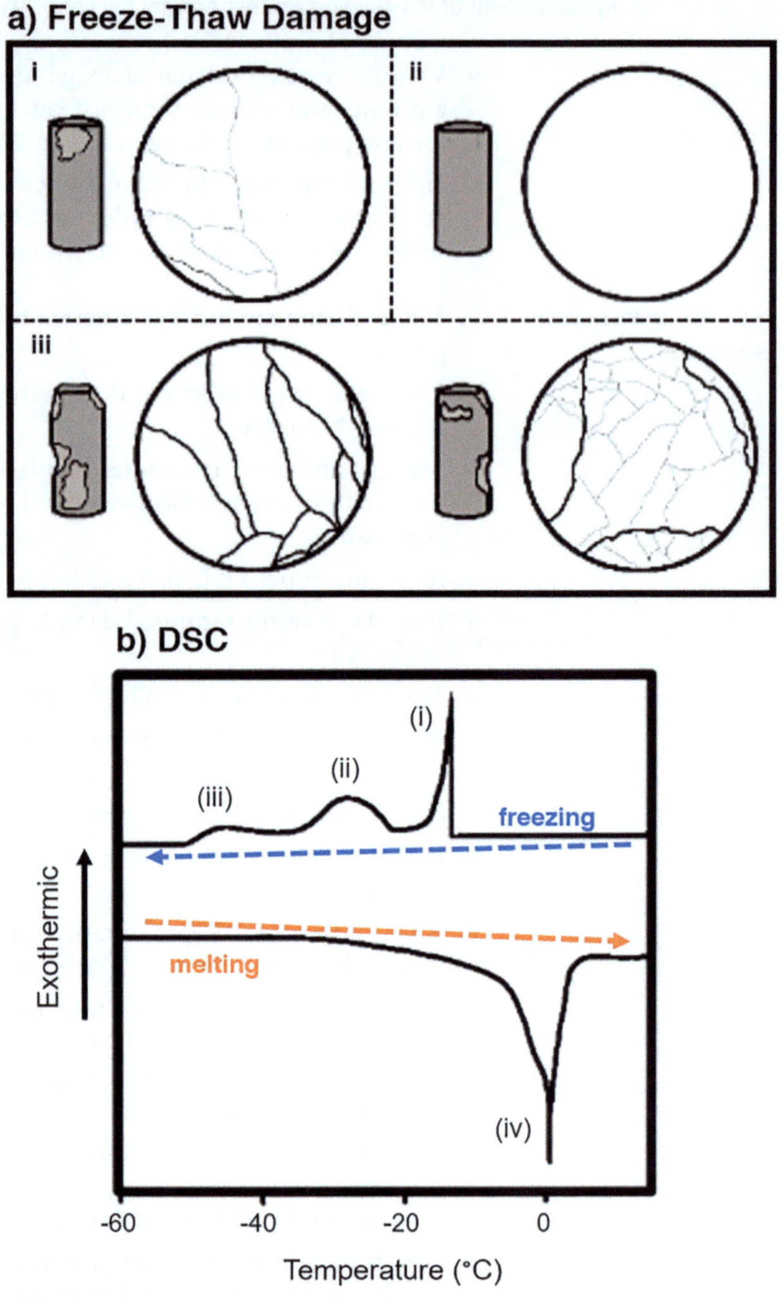

Fig. 2 Examples of data for determining IBP performance within cement paste. (**a**) Examples of macroscopic samples and microscopic cross sections of cement cylinders for freeze-thaw damage (adapted from Frazier et al. [8]). (i) Degradation for a control sample, with some macroscopic surface scaling and microscopic cracks. (ii) Degradation for an active sample, with little to no macroscopic surface scaling or microscopic cracks. (iii) Two examples of degradation for an inactive sample, with a lot of macroscopic surface scaling and microscopic cracks. (**b**) DSC data that demonstrates the formation of ice within cement paste. (i) Super-cooling freezing of macropores. (ii) Freezing of mesopores (sometimes called capillary pores). (iii) Freezing of micropores (sometimes called gel pores). (iv) Melting/thawing of all ice

4 Notes

1. PBS cannot be used for complex alkaline solutions. The divalent cations in complex alkaline solutions form salt precipitates with the phosphates in PBS.
2. IBPs should be incubated at ambient temperature in each solution for 24 h prior to analysis to ensure an equilibrium folding state has been reached.
3. At least two detectors should be used to verify IBP elution peak shape and time.
4. Similar columns may be used instead, as long as the column specifications are appropriate for the IBP being investigated. It is important to look at the size range for separation, solution tolerance, and flow rate tolerance.
5. Running buffer should closely match the constituents of the sample buffer. The running buffer should be neutral as per equipment and column specifications, as most cannot handle alkaline pH. For best results, match the cations of the sample buffer by using chloride salts, e.g., create similar Na^+ concentrations by using NaCl instead of NaOH.
6. The volume required is dependent on the injection method for SEC. For manual injection, at least 50–100 μL of solution is required. For typical autosamplers, 0.5 mL is necessary for proper solution uptake. However, small volume inserts (sometimes called squid feet) are available and reduce the needed volume for autosamplers to only 100 μL.
7. 1 mg/mL is ideal for prominent SEC peaks; however, 0.5 mg/mL can be sufficient if the noise from the solution is not too large. A general rule to keep in mind is that lower molecular weight proteins require a higher concentration for analysis.
8. 0.5 mg/mL is necessary for observing impurities and potential aggregation or degradation. However, 0.5 mg/mL might "blow out" the band on the gel, and the concentration may need to be reduced for a publishable gel. 0.2 mg/mL has been found to be an informative and attractive concentration for publishable gels.
9. Lab-poured gels are an acceptable alternative to pre-poured gels, especially for preliminary results. However, a pre-poured gel is recommended for manuscript images.
10. IBPs should have a concentration of 0.5 mg/mL to ensure collection of sufficient absorbance spectra; however, 0.2 mg/mL may be sufficient to account for dilutions to reduce salt content.

11. Large quantities (≥100 mg) of protein are required for analysis within cement paste. It is recommended to use IBPs that demonstrate exceptional potential in alkaline solutions or IBPs that are easily obtained in large quantities.
12. Other cement can be used as long as it complies with ASTM C192 [15].
13. The cement paste should be evenly dispersed along the bottom of the container for thorough mixing.
14. Cement paste is difficult to remove from glassware, so dedicated reusable containers or disposable containers are recommended to prevent lab-wide contamination.
15. Other cylinder sizes may be used as long as they comply with ASTM C39 [16].
16. The container for sodium phosphate should fit inside the airtight sealable container.
17. The freeze-thaw chamber may be built or purchased. It should have a temperature range that includes −15 °C and 16.5 °C (± 2 °C) and the ability to set time points between 1 and 3 h and allow at least 30 cycles.
18. Not all freeze-thaw chambers require ice. Some purchased chambers use a Peltier plate system; however, some built systems require ice-cooled recirculation.
19. XRM is a cost-prohibitive analysis step. A large working distance microscope may be able to be used instead to provide some characterization of internal damage.
20. The DSC should be able to reach −90 °C with a ramp rate of 5 °C/minute.
21. pH may vary between cold and ambient temperatures. To determine the difference, the sample pH can be recorded while the sample is placed in an ice bath. When considering applications in cement, majority of the time, the IBP will be at ambient temperature.
22. It is important to test multiple alkaline solutions around the pH of concrete (e.g., 10, 11, 12, 13) to determine the range of IBP stability.
23. While concrete pore solution can reach pH > 13, salt precipitation is likely at pH > 13.2, which can be detrimental to equipment. Thus, pH 13 is a recommended upper bound.
24. Alkaline solutions with divalent cations have the possibility of creating metal complexation gels with high molecular weight molecules (>90 kDa). It may be beneficial to test an aliquot of sample for potential gelation (allow 24+ hours to observe).

25. The IBP will precipitate out of solution around its isoelectric point, making it difficult to characterize. It is recommended to test a pH ~0.3 above or below the isoelectric point to avoid potentially damaging equipment. If a pH around the isoelectric point must be used, the solution should be centrifuged and the supernatant should be characterized.
26. The amount of solution to incorporate will depend on the concentration of IBP you are able to obtain during purification. It is recommended that the total IBP in alkaline solution is at least 1 mg/mL, but 2 mg/mL is better.
27. Instead of pre-weighing NaOH, add as a liquid while monitoring using a pH meter. The incorporated IBP will influence the final pH of the solution, and the amount of NaOH needed for the target pH will depend on which IBP is being included.
28. $Ca(OH)_2$ should always be added as a powder to ensure saturation. After mixing, solutions should be centrifuged to remove excess $Ca(OH)_2$. The supernatant should be used for further testing.
29. For pH 13 solution, 5 M NaOH should be used to reduce the volume added, switching to 1 M NaOH when it is within ~0.3 and then switching to 0.1 M NaOH when it is within ~0.05.
30. If the pH goes above the intended value, adjust using 0.1 M HCl. For pH 13 solutions, 1 M HCl may be required to adjust the pH.
31. When making control solutions, add the same constituents that would be incorporated from the IBP solution (except for the IBP), e.g., if the IBP solution had Tris in it—include Tris in the control solution. Similarly, the same amount of pH adjustor should be added. The pH may be slightly different due to the influence of the IBP on the pH.
32. Small deviations in data are expected due to the influence of pH adjustors and can be mostly disregarded. Examples of acceptable deviations are demonstrated in Fig. 1.
33. Running buffer should be made at most 24 h before equilibrating.
34. The time required for equilibration is system dependent (typically between 2 and 12 h). A good rule of thumb is to purge ~3× the volume of the system before recycling solvent.
35. Centrifugation is especially important for highly alkaline, highly ionic samples as salts and precipitate are very damaging to the SEC column.
36. Due to the small pore size of the filter, pressure should be applied slowly to prevent popping the filter off the syringe. It takes a few minutes to filter ≤500 μL of sample. Some IBPs are

not able to be filtered, as the sheer force may damage the structural integrity of the IBP. An IBP in its preferred solution should be tested after filtration to ensure filtration does not cause degradation.

37. 30 min is often sufficient for IBP elution. However, a run time of 45 min ensures that no sample will remain in the system or detectors to prevent potential contamination.
38. 0.185 is an average dn/dc for proteins and should suffice. If you have determined a more specific dn/dc, input that instead.
39. Due to the pH difference between samples to the running buffer, purging the equipment for ~1 h between uses can improve spectra noise.
40. Since the sample solution will not match the running solution due to the pH adjustors, there will be significant noise in detectors (especially RI, and somewhat with UV). MALS or viscosity is best for reducing the noise to detection ratios, although RI and UV will still provide usable, but noisier, data.
41. Test for leaks in the electrophoresis setup. If it leaks, drain, disassemble, and reassemble. Repeat until it does not leak.
42. Protein ladder and samples should be deposited slowly into the gel wells to prevent ejection into the running buffer solution.
43. The amount of protein ladder and sample to load into wells depends on the size of the well. Gels with more wells will need less, and gels with fewer wells will need more.
44. Check on the gel every 10–15 min. Sometimes the same setup will vary in time.
45. If the gel background stays a persistent blue, it can be cleaned further by using a 20% NaCl solution instead of Milli-Q.
46. pH adjustors absorb in the peptide bond region, so each alkaline solution should be tested to determine respective dilutions to prevent detector saturation.
47. Spectra should exhibit a lower wavelength limit between 180 and 190 nm to allow deconvolution. Alkaline solutions may have slightly different lower wavelength limits.
48. The Na^+ concentration for pH 13 solutions is often too high to collect enough data for deconvolution without drastically reducing the pH. However, some spectra can be collected to form a hypothesis of IBP folding.
49. The pH of the solution may drop by ~0.2–0.4. This is unavoidable due to the necessary dilution to prevent confluence of the solution constituents with the detector, and insights on IBP conformation around that pH can still be provided.

50. Make sure that the cuvettes are cleaned and dried in between every test—residual pH adjustor or IBPs will cause interference with the new absorbance spectra.

51. IBPs around their isoelectric point can destroy a CD cuvette, drastically increasing the noise in addition to greatly reducing IBP absorbance spectra. If possible, investigate at a pH $\pm$ 0.3 from the isoelectric point.

52. Simple alkaline solutions are readily investigated using the splat method. Complex alkaline solutions should be investigated using the sandwich method as their high viscosity prevents the formation of an ice monolayer using splats, and $Ca(OH)_2$ precipitate acts as nucleation seeds.

53. Use the control alkaline solution for serial IBP dilutions to ensure the solution constituents are the same and only the concentration of IBP is changing.

54. If you look down the inside of the PVC pipe, you should be able to see the glass slides. Positioning the slides this way will facilitate splat accuracy.

55. Warmer weather needs increased cooling time (e.g., 5 min in winter, 15 min in summer).

56. A monolayer for a good splat will look like well-defined grains of rice. A monolayer for a good sandwich will look like a lot of small "bubbles" with some, but preferably minimal, impingement.

57. Solution constituents can influence recrystallization dynamics, which could obfuscate the influence of the IBP.

58. After 3 min of stirring, the cement mix should have the consistency of maple syrup.

59. To achieve minimal air bubbles, approximately 20–30 taps are required. The same number of taps should be used across sample types.

60. The slurry is used to create an ~99% relative humidity condition for cement curing. Slurry creation should be done according to ASTM E104 [17].

61. Periodically check container within the 14-day curing period to make sure there is condensation on the lids/sides of the container. If there is no condensation, repeat **step 8**.

62. There should be enough room so the samples do not touch other samples.

63. Throughout cycling, samples should be "shuffled" so that they do not all remain in the same spots within the container.

64. Three potential sample pieces should be generated to ensure that at least one will appropriately fit into the DSC pan. It is important that the bottom of the sample lay flat on the DSC pan and cover most of the bottom of the pan.
65. The water level in the scintillation vial should completely cover the DSC samples.

Acknowledgments

This review was made possible by the Department of Civil, Environmental, and Architectural Engineering, the College of Engineering and Applied Sciences, and the Living Materials Lab at the University of Colorado Boulder with financial support from the US National Science Foundation (award no. CMMI-1727788). This work represents the views of the authors and not necessarily those of the sponsors.

References

1. Delesky EA, Thomas PE, Charrier M et al (2020) Effect of pH on the activity of ice-binding protein from Marinomonas primoryensis. Extremophiles 25(1):1–13
2. Delesky EA, Frazier SD, Wallat JD et al (2019) Ice-binding protein from Shewanella frigidimarinas inhibits ice crystal growth in highly alkaline solutions. Polymers 11(2):9–12
3. Ptitsyn OB (1987) Protein folding: hypotheses and experiments. J Protein Chem 6(4): 273–293
4. Berg JM (2002) Biochemistry 5th ed.
5. Gambhir ML (2013) Concrete technology: theory and practice. Tata McGraw-Hill Education
6. Scherer GW, Valenza JJ (2005) Mechanisms of frost damage. Mater Sci Concr 7(60):209–246
7. Scherer GW (1999) Crystallization in pores. Cem Concr Res 29(8):1347–1358
8. Ghods P, Isgor OB, McRae G et al (2009) The effect of concrete pore solution composition on the quality of passive oxide films on black steel reinforcement. Cem Concr Compos 31(1):2–11
9. Knight CA, Wen D, Laursen RA (1995) Nonequilibrium antifreeze peptides and the recrystallization of ice. Cryobiology 32(1):23–34
10. Goñi S, Andrade C (1990) Synthetic concrete pore solution chemistry and rebar corrosion rate in the presence of chlorides. Cem Concr Res 20(4):525–539
11. Moragues A, Macias A, Andrade C (1987) Equilibria of the chemical composition of the concrete pore solution. Part I: comparative study of synthetic and extracted solutions. Cem Concr Res 17(2):173–182
12. Knight CA, Hallett J, Devries AL (1988) Solute effects on ice recrystallization: an assessment technique. Cryobiology 25:55–60
13. Olijve LC, Oude-Vrielink AS, Voets IK (2016) A simple and quantitative method to evaluate ice recrystallization kinetics using the circle Hough transform algorithm. Cryst Growth Des 16(8):4190–4195
14. Wu S, Zhu C, He Z et al (2017) Ion-specific ice recrystallization provides a facile approach for the fabrication of porous materials. Nat Commun 8:1–8
15. ASTM Standard C192 (2002) Standard Practice for Making and Curing Concrete Test Specimens in the Laboratory
16. ASTM Standard C39 (2003) Standard Test Method for Compressive Strength of Cylindrical Concrete Specimens
17. ASTM Standard E104 (2005) Standard Practice for Maintaining Constant Relative Humidity by Means of Aqueous Solutions

Chapter 11

Measurement of Ice-Binding Protein Inhibition of Non-ice Crystal Growth

Michihiro Muraoka

Abstract

The kinetic hydrate inhibitor (KHI) was developed to prevent the formation of undesirable gas hydrate crystals in natural gas pipelines. Studies of antifreeze proteins (AFPs) are gaining attention in the natural gas research field due to their performance in crystal growth inhibition, excellent biodegradation, and low toxicity. Studies of AFPs may provide clues for developing future commercial KHIs used offshore. This chapter presents a simple method of evaluating AFP inhibitory performance as a KHI on tetrahydrofuran (THF) hydrate growth with a unidirectional growth apparatus.

Key words Ice-binding proteins, IBP, Antifreeze proteins, AFP, Non-ice crystal growth, Tetrahydrofuran hydrate, THF hydrate, Gas hydrate, GH, Unidirectional growth apparatus

1 Introduction

Type I antifreeze proteins (AFPs) were initially found in the blood plasma of a fish known as the winter flounder [1]. AFPs have a simple α-helical structure [2]. AFPs adsorb on the bipyramidal plane surface of ice crystals at low concentrations [3]. AFPs adsorb on all orientation surfaces of ice crystals at high concentration.

AFP type III was first found in an ocean pout [4]. It has a globular protein [5]. AFP type III is absorbed on the prism and pyramidal planes of ice crystals [6]. This AFP technology is gaining a lot of attention in the field of natural gas recovery engineering for the following reasons:

A natural gas hydrate is an ice-like crystal that incorporates guest gas molecules within hydrogen-bonded water cages. A gas hydrate forms either structure one (sI) or structure two (sII) clathrate hydrate under low temperatures and high pressure [7]. Large amounts of naturally formed methane hydrates are found in the

Ran Drori and Corey A. Stevens (eds.), *Ice Binding Proteins: Methods and Protocols*, Methods in Molecular Biology, vol. 2730, https://doi.org/10.1007/978-1-0716-3503-2_11,

sediment under the ocean floor and in permafrost regions. These methane hydrates could be an interesting source of energy in the future.

In 1934, Hammerschmidt discovered gas hydrate blockages in pipelines [8]. The formation of natural gas hydrates in pipelines and production equipment is a serious problem. A thermodynamic hydrate inhibitor (THI), or kinetic hydrate inhibitor (KHI), has been developed to prevent gas hydrate formation in high-pressure natural gas pipelines. THI shifts the phase equilibrium condition to a higher pressure and lower temperature. To prevent gas hydrate formation in operation site, typical THIs such as methanol or ethylene glycol require 20–50 wt% to be added in water. KHIs are generally water-soluble polymers such as polyvinylpyrrolidone (PVP) or polyvinylcaprolactam (PVCap). The KHIs primarily act to delay gas hydrate nucleation and inhibit gas hydrate growth rate. KHI requires only 0.1–1.0 wt% to be added to water [9]. KHIs have advantages such as low transportation costs and minimal impact on the environment surrounding the operation site.

For preventing environmental problems, the following three key ecotoxicological tests are most important for chemicals containing KHIs used offshore. These are toxicity, bioaccumulation, and seawater biodegradation [10]. PVP and PVCap are major KHIs in commercial use, but they are poorly biodegradable. New KHI products should have an environmental rating of yellow 1 [11]. The environmental rating definition of green, yellow, red, and black in Norway is described in ref. [11]. The new KHIs are required to be over 60% biodegradable, as measured by the OECD306 seawater test in 28 days [12, 13]. That is a tough target for new KHIs in addition to having high performance and remaining cost-efficient.

Experiments on gas hydrate formation require high-pressure gas experiments, such as those performed at a pressure of 10 MPa. The tetrahydrofuran (THF)–water system is a useful gas hydrate formation experiment model. THF is miscible with water at all molar ratios, and a stoichiometric THF–water solution (mole ratio of THF/H_2O = 1:17) forms a THF hydrate crystal (sII) under atmospheric pressure below 4.4 °C [14].

Zeng et al. show that AFP type I inhibits the growth of THF hydrate [15]. AFP type I also affects the morphology and inhibits the nucleation of THF hydrate crystals. Zeng et al. show that AFP type I inhibits the growth of methane and propane hydrate [16]. Additionally, AFP type I inhibits the nucleation of the hydrates.

Daraboina et al. studied the inhibition effect of AFP type III on gas hydrate. The results of the study showed that AFP type III delayed the onset of hydrate nucleation [17]. The results also showed that AFP type I was more efficient than AFP type III.

In the early stages of KHI technology development, researchers were inspired by the studies of AFP inhibition function on ice. In recent years, the AFP studies have attracted attention from KHI researchers for their excellent environmental compatibility, such as biodegradation. Studies of AFPs may provide clues for developing future commercial KHIs used offshore.

The exact AFP and KHI inhibition mechanisms on ice and gas hydrate are still not clear. Some studies suggest that the Gibbs–Thomson effect can explain the AFP and KHI inhibition mechanisms on ice and gas hydrates [1, 2, 18–21]. We believe that both the AFP and KHI inhibition mechanisms on ice and gas hydrate are very similar and closely related.

In most of the KHI inhibition effect evaluation experiments, gas hydrates are formed by stirring gas and water in a high-pressure reactor [22, 23], or THF hydrates are formed in a rocking cell [24, 25]. It is important that these types of experimental systems separate the inhibition effect of crystal nucleation and growth. In conclusion, a simplified experimental data is needed to clarify the KHI inhibition mechanism.

This chapter presents a simple technique for evaluating AFP inhibition activity as a KHI on THF hydrate growth with a unidirectional growth apparatus. The technique can only evaluate the effect of KHIs on growth inhibition. Please note that this technique cannot evaluate the effect of nucleation rate inhibition. We developed this technique to clarify the KHI growth inhibition mechanism and propose a unified index for KHI growth inhibition performance on THF hydrate crystals. Our research progress is published in Refs. [20, 21, 26, 27]. We summarized our own method procedure in Sects. 2 and 3 with important notes in Sect. 4. If you are making the same experimental setup, be sure to read **Note 3**.

2 Materials

The following materials are needed for preparing sample solutions:

1. Prepare ultra-pure water (18.2 MΩ × cm resistivity).
2. Prepare dehydrated stabilizer-free THF (99.5 wt% purity, supplied by Kanto Chemical Co., Japan).
3. Prepare high-purity AFP type I extracted from the winter flounder (average molecular weight = 3200, Nichirei Co., Japan).
4. Prepare high-purity AFP type III from the ocean pout or eel pout (average molecular weight = 6700, Nichirei Co., Japan).

3 Methods

3.1 How to Make an Experimental Cell

How to make an experimental cell is shown in Fig. 1. Prepare the glass plates (26 × 76 × 1 mm, crown grass or quartz), the rubber sheet (152.4 × 152.4 × 0.51 mm, Kalrez 6375, 5022, DuPont Co.), and the thermocouples (wire thickness = 0.10 mm, type T, 01-T, Ninomiya Electric Wire Co. Japan). We recommend the quartz grass for its high thermal conductivity performance. Note that the rubber sheet needs THF resistance. The erosive power of THF on the normal chemical resistance of rubber containing fluor-elastomers and silicone rubber is very strong.

1. We use the spot-welding machine (HSW-03, Yokodai.JP Co., Japan) to weld copper and a constantan wire of thermocouple.
2. The cut sizes of strips from rubber sheet are 70 × 2 mm and 7 × 2 mm.
3. Place the rubber strips on the bottom of the glass plate.
4. Place the thermocouples on the rubber strips. The thermocouples are located about 20 mm from the right inner edge of the blocks.

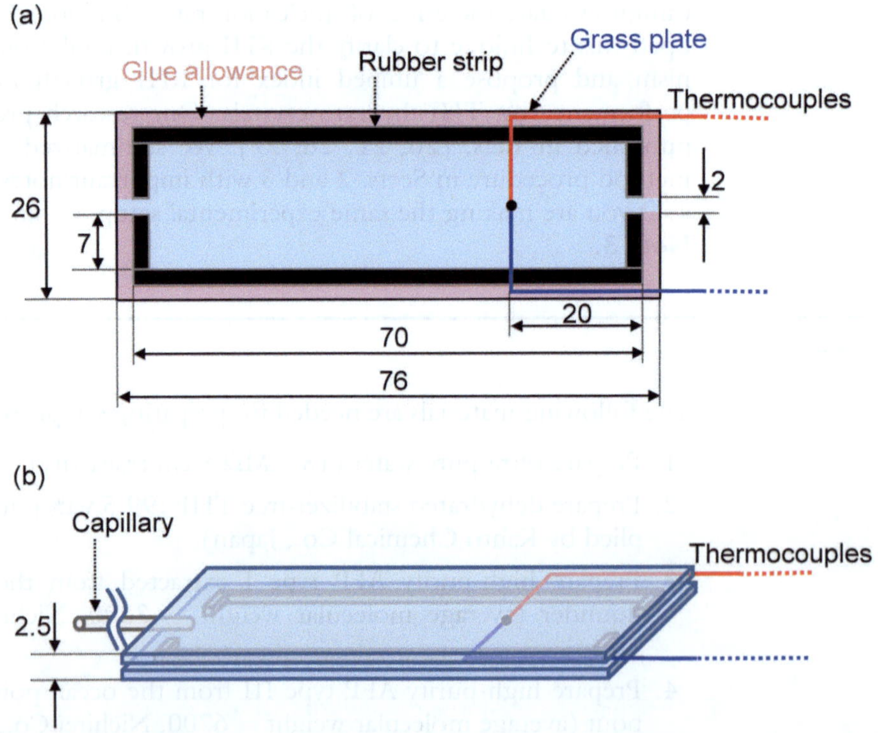

Fig. 1 **(a)** Top view of a cell assembly diagram. The black parts are rubber strips. The pink-colored parts are for glue allowances. **(b)** A diagram of the completed cell

5. Place the top side glass plate on the rubber strips and thermocouples.
6. Pinch the cell from the top and bottom.
7. Apply epoxy bonding agent (Araldite Rapid, Huntsman Co.) to the glue allowance on both long sides of the cell.
8. Allow the adhesive to cure for 24 h.
9. Insert the capillary (PFA microtube, inner diameter × outer diameter = 0.4 × 0.6 mm, AS ONE Co., Japan) into the outlet side of the cell.
10. Apply epoxy bonding agent to the glue allowance on both short sides of the cell.
11. Allow the adhesive to cure for 24 h.
12. Caliper the thickness of the completed cell. A range of acceptable thickness is 2.5–2.6 mm.

3.2 Methods of Preparing THF–Water

A stoichiometric THF–water solution with a mole ratio of THF/H_2O = 1:17 was prepared.

1. Vacuum the ultra-pure water with a vacuum pump and magnetic stiller for 10 min.
2. Vacuum the THF with a chemical vacuum pump (THF resistance) and magnetic stiller for 5 min.
3. Pour 76.565 g of water and 18.030–18.050 g of THF into a bottle (*see* **Note 1**).
4. Shake the bottle 100 times to make sure the solution is well mixed.
5. Place the bottle in an ultrasonic cleaner and agitate it with ultrasonic waves for 10 min. Do this while cooling the water in the cleaner with ice.

3.3 Methods of Preparing Sample Solutions

Add the AFP to the THF–water solution. The following is an example of a weight concentration $c = 0.5$ wt%.

1. Weigh the 0.53 g AFP.
2. Pour the AFP into an empty bottle.
3. Pour the 10.00 g of THF–water solution into the bottle.
4. Mix the sample solution with a magnetic stirrer until the AFP is completely dissolved.

3.4 Methods of Preparing Sample Cell

1. Collect 1 mL of sample solution with a syringe.
2. Inject the sample solution into the cell from the inlet port.

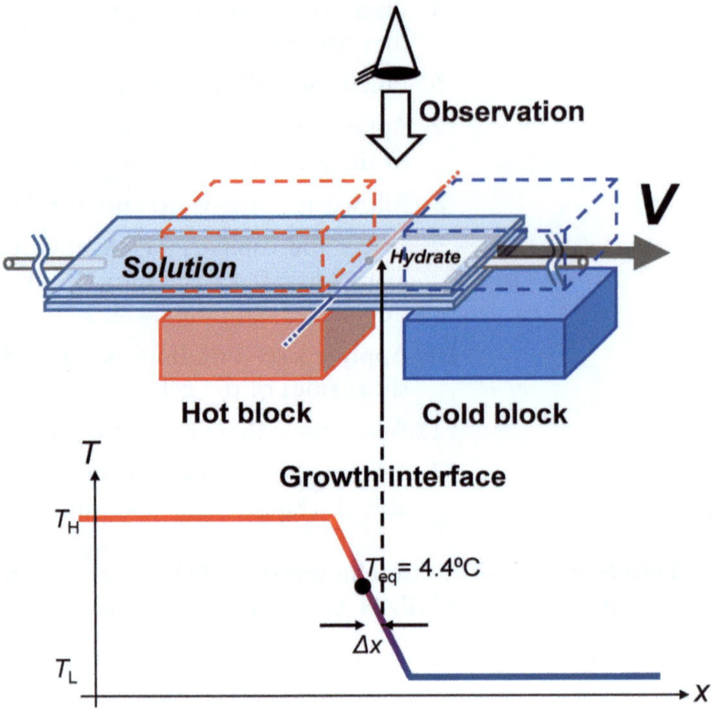

Fig. 2 A schematic of the unidirectional growth apparatus. This figure is reused from Fig. 1 in Ref. [21]

3.5 Information About the Unidirectional Growth Apparatus

The unidirectional growth apparatus is shown in Fig. 2. The sizes of copper blocks are shown in Fig. 3. The hot and cold blocks were made of copper. The distance between the hot and cold blocks is 5 mm. This 5 mm distance is the observable area by the lens and camera. A constant temperature gradient, *G*, is applied at the 5 mm distance. The surface of the blocks is polished to a mirror-like finish. The temperature of the blocks was controlled by thermoelectric modules. The modules are comprised of Peltier devices (39.7 × 39.7 × 3.94 mm, I max 8.5 A, V max 17.5 V, and 9500/127/085B, Ferro Ted Co.) and a Peltier controller (TDC-1020A, Cell System Co., Japan). The exhaust heat from the Peltier device is provided by the circulation of refrigerant from the chiller (CTP-3000, EYELA Co., Japan) (*see* **Note 2**). The temperature of the refrigerant is set at 1 °C (*see* **Note 3**). The platinum resistance thermometers (Pt 100, three-wire type, outer diameter of protection tube = 2.0 mm, length of protection tube = 23 mm, R620-1, Chino Co., Japan) are inserted into the holes for Pt thermometer of hot and cold blocks as shown in Fig. 3a.

Our motor system was composed of a pulse motor (PK543BW-H100S, basic step angle = 0.0072°), motor driver (CRD5107P), and programmable controller (EMP401-1). All motor components

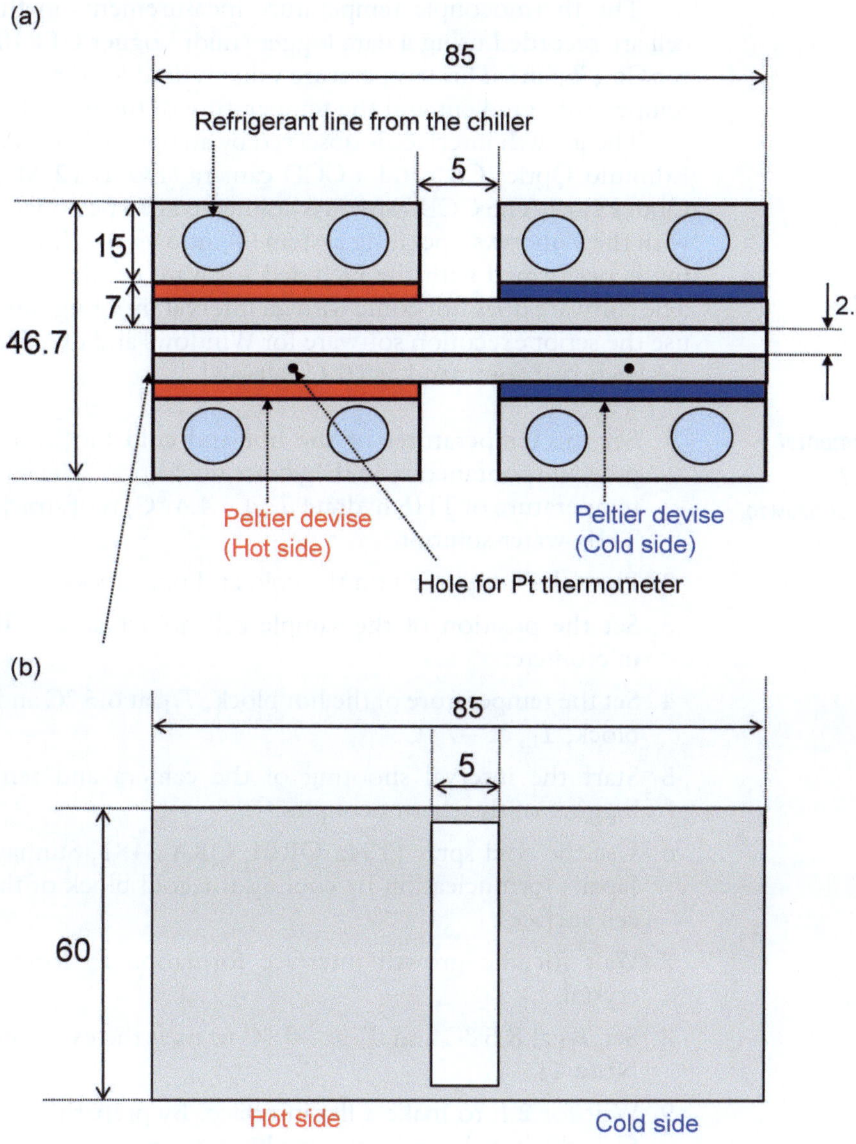

Fig. 3 Size information of the copper blocks. (**a**) A diagram of copper blocks from the side view. (**b**) A top view of a surface copper plate

were purchased from Oriental Motor Co., Ltd., Japan. This motor is connected to the micrometer head part of the x-axis stage (limits of movement = 25 mm, movement per rotation = 0.5 mm, TSD-65171S-M6, SIGMAKOKI Co., Japan). The minimum velocity V in this system is 0.0004 μm s^{-1}. The maximum velocity, V, is more than 100 μm s^{-1}. The x-axis stage is connected to the sample cell holder and used for the sample cell movement.

The thermocouple temperature measurements in the sample cell are recorded using a data logger (midi Logger GL840, Graphtec Co., Japan). This temperature value is used for determining the temperature gradient and the temperature of the growth interface.

The growth interface is observed by an optical lens (VZM 600i, Edmund Optics Co.) and a CCD camera (EO-1312 M, Edmund Optics Co.). The CCD camera is connected to a personal computer with the Windows operating system (Windows 10). Interval shooting is performed with the included software of the CCD camera. The software does not come with an interval shooting function. We use the script execution software for Windows and execute the *Ctrl* + *S* keyboard command at 10 s intervals.

3.6 Experimental Procedure of Unidirectional Growth

1. Set the temperatures of the hot and cold blocks at 5 °C to prevent spontaneous THF hydrate nucleation. The equilibrium temperature of THF hydrate T_{eq} (= 4.4 °C) is of stoichiometric THF–water solution.
2. Place the sample cell on the cold and hot blocks.
3. Set the position of the sample cell holder at $x = 0$ mm by micrometer.
4. Set the temperature of the hot block, T_H, at 6.5 °C and the cold block, T_L, at −7 °C.
5. Start the interval shooting of the camera and temperature logging of the thermocouples.
6. Use the cold spray (134a QREI, QRA-S481, Sunhayato Co., Japan) for nucleation by cooling the cold block of the sample cell surface.
7. Wait for the growth interface formation to form an extra crystal.
8. Set T_H at 8.5 °C and T_L at −7 °C to melt the excess crystal (*see* **Note 4**).
9. Wait for 2 h to make a flat interface. By preparing a flat interface, the initial conditions can be standardized.
10. Move the pulse motor at a constant velocity toward the cold block.

3.7 Example Pictures of THF Hydrate Growth Interface Shift

The sequential images of a THF hydrate growth interface shift (c of AFP $I = 0.5$ wt% and $V = 5$ μm s^{-1}) are shown in Fig. 4. The recorded times are (a) $t = 0$ min, (b) $t = 3.3$ min, (c) $t = 16.2$ min, and (d) $t = 67$ min. At $t = 67$ min (*see* Fig. 4d), the growth interface reached the thermocouple tip. The degree of supercooling of the growth interface, ΔT, is determined by the temperature value measured by the thermocouple when it encounters the growth interface.

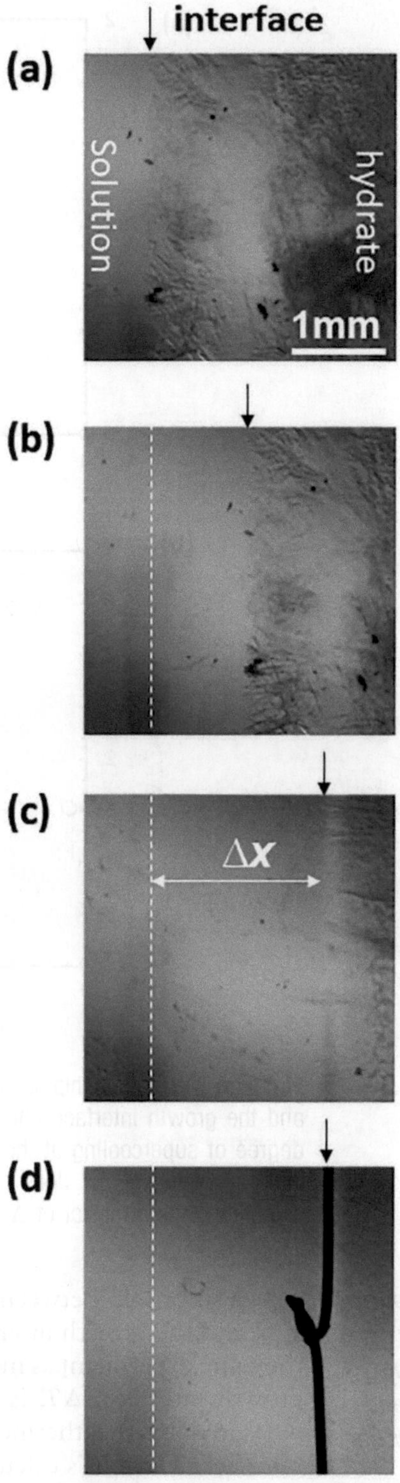

Fig. 4 The sequential images of a THF hydrate growth interface shift at (**a**) $t = 0$ min, (**b**) $t = 3.3$ min, (**c**) $t = 16.2$ min, and (**d**) $t = 67$ min

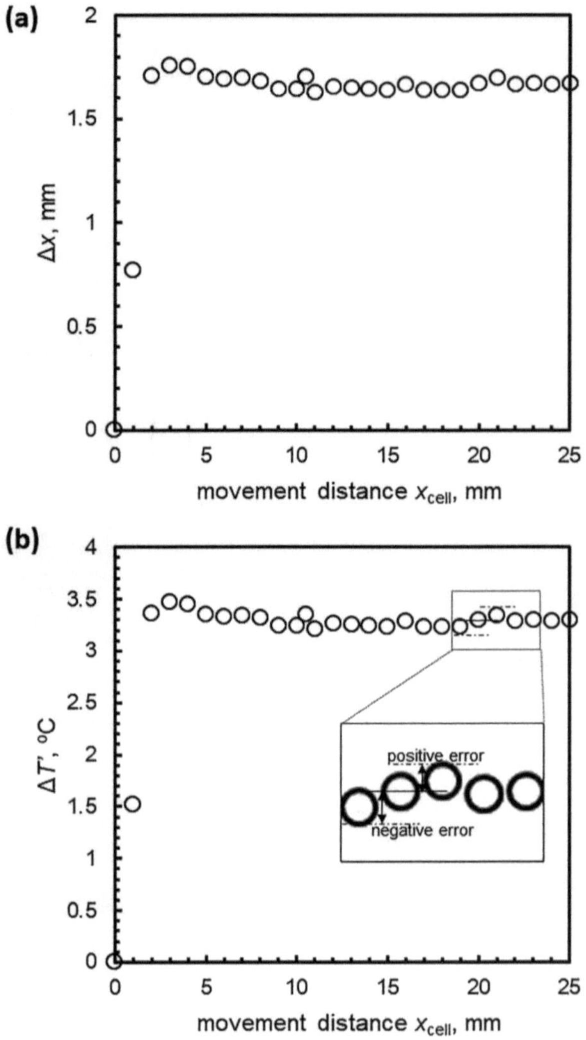

Fig. 5 (**a**) The relationship between the movement distance of a sample cell, x_{cell}, and the growth interface shift, Δx. (**b**) The relationship between x_{cell} and the degree of supercooling at the growth interface, $\Delta T'$, from $G\Delta x$. Note that the accurate value of ΔT does not agree with $\Delta T'$. This $\Delta T'$ value is used for calculating the error bar of ΔT

3.8 How to Measure the Degree of Supercooling of the Growth Interface and Determine Its Error Bar

The relationship between the movement distance of a sample cell, x_{cell}, and the growth interface shift, Δx, is shown in Fig. 5a. This is the same experiment as in Fig. 4. The degree of supercooling of the growth interface, ΔT, is determined by the temperature T value measured by the thermocouple when it encounters the growth interface. The ΔT is calculated as follows:

$$\Delta T = 4.4\,^{\circ}\mathrm{C} - T. \quad (1)$$

From the experiment, $\Delta T = 2.4$ °C. The value of ΔT shows the inhibition activity of AFPs on the THF hydrate crystal growth. Note that this measurement requires conducting under steady-state condition. In this experiment, the growth interface shift reached a steady-state condition after $x_{cell} = 5$ mm.

The relationship between x_{cell} and the degree of supercooling at the growth interface, $\Delta T'$, is shown in Fig. 5b. $\Delta T'$ is calculated as follows:

$$\Delta T' = G\Delta x \quad (2)$$

The G is the measured temperature gradient between the cold and hot blocks (observable area). Note that the accurate degree of supercooling of the growth interface cannot be determined from the $\Delta T'$ value because the temperature of the initial growth interface does not reach the equilibrium temperature due to the AFPs' effect [27]. We use this $\Delta T'$ value in Fig. 5b for evaluating the error bar. The error bar is given by the $\Delta T'$ fluctuation from plots around $x_{cell} = 20$ mm. When the thermocouples touch the growth interface, $x_{cell} = 20$ mm. The positive and negative errors of ΔT are given as shown in Fig. 5b. As a result, $\Delta T = 2.4 \pm 0.1$ °C is given by this experiment. The results of our entire study on AFP types I and III are summarized in ref. [27].

4 Notes

1. The tip of the stoichiometric THF–water solution experiment is adding one more drop of THF than the stoichiometric composition. This is because THF volatilizes a little bit during the operation of the experiment. For example, we mix 18.03 or 18.04 g of THF with 76.565 g of water.
2. The temperature fluctuation of the chiller's refrigerant directly becomes the temperature fluctuation of the temperature-controlled copper block. A chiller with a heater function as well as cooling is recommended. We chose a chiller (CTP-3000, EYELA Co., Japan) with a temperature stability of less than 0.1 °C.
3. If the Peltier element operates without cooling water circulating, abnormal heat generation will occur. This is very dangerous as it can cause a fire. We use the following failsafe system. When the cooling water is not circulating, the Peltier controller is turned off by a flow switch. We also use the failsafe system of the temperature limiter of the Peltier controller module. The Peltier is turned off when the temperature is over 30 °C by this limiter. At least a double failsafe system is strongly recommended.

4. The appropriate temperature setting of T_H and T_L differs depending on experimental conditions such as AFP type and concentration. Please think flexibly and decide the temperature setting values. In addition, the position of the thermocouples in the cell can be flexible for experimental purposes.

Acknowledgments

AFP types I and III are currently (2021) supplied by Nichirei Co., Japan, for testing and research purposes only.

References

1. Raymond JA, DeVries AL (1977) Adsorption inhibition as a mechanism of freezing resistance in polar fishes. Proc Natl Acad Sci U S A 74(6): 2589–2593
2. Knight CA (2000) Adding to the antifreeze agenda. Nature 406(6793):249–251
3. Furukawa Y, Inohara N, Yokoyama E (2005) Growth patterns and interfacial kinetic supercooling at ice/water interfaces at which antifreeze glycoprotein molecules are adsorbed. J Cryst Growth 275(1-2):167–174
4. Nagashima K, Furukawa Y (1997) Nonequilibrium effect of anisotropic interface kinetics on the directional growth of ice crystals. J Cryst Growth 171(3-4):577–585
5. Zepeda S, Uda Y, Furukawa Y (2008) Directly probing the antifreeze protein kinetics at the ice/solution interface (< special issue> crystal growth controlled by macromolecules). J Jpn Assoc Cryst Growth 35(3):151–160
6. Antson AA, Smith DJ, Roper DI, Lewis S, Caves LS, Verma CS, Buckley SL, Lillford PJ, Hubbard RE (2001) Understanding the mechanism of ice binding by type III antifreeze proteins. J Mol Biol 305(4):875–889
7. Sloan ED Jr, Koh CA (2008) Clathrate hydrates of natural gases, 3rd edn. CRC Press, Boca Raton
8. Hammerschmidt EG (1934) Formation of gas hydrates in natural gas transmission lines. Ind Eng Chem 26(8):851–855
9. Kelland MA (2006) History of the development of low dosage hydrate inhibitors. Energy Fuel 20(3):825–847
10. Kelland MA (2018) A review of kinetic hydrate inhibitors from an environmental perspective. Energy Fuel 32(12):12001–12012
11. Norwegian Oil and Gas Association (2019) Recommended guidelines for emission and discharge reporting. The Norwegian Oil and Gas Association, Stavanger
12. Kelland MA (2014) Production chemicals for the oil and gas industry, 2nd edn. CRC Press, Boca Raton. https://doi.org/10.1201/b16648
13. Organization for Economic Co-operation and Development (OECD) (2002) OECD guideline for testing of chemicals: biodegradability in seawater. Paris, OECD, p 27
14. Gough SR, Davidson DW (1971) Composition of tetrahydrofuran hydrate and the effect of pressure on the decomposition. Can J Chem 49(16):2691–2699
15. Zeng H, Wilson LD, Walker VK, Ripmeester JA (2006) Effect of antifreeze proteins on the nucleation, growth, and the memory effect during tetrahydrofuran clathrate hydrate formation. J Am Chem Soc 128(9):2844–2850
16. Zeng H, Moudrakovski IL, Ripmeester JA, Walker VK (2006) Effect of antifreeze protein on nucleation, growth and memory of gas hydrates. AIChE J 52(9):3304–3309
17. Daraboina N, Ripmeester J, Walker VK, Englezos P (2011) Natural gas hydrate formation and decomposition in the presence of kinetic inhibitors. 1. High pressure calorimetry. Energy Fuel 25(10):4392–4397
18. Nada H, Furukawa Y (2012) Antifreeze proteins: computer simulation studies on the mechanism of ice growth inhibition. Polym J 44(7):690–698
19. Yagasaki T, Matsumoto M, Tanaka H (2018) Molecular dynamics study of kinetic hydrate inhibitors: the optimal inhibitor size and effect of guest species. J Phys Chem C 123(3): 1806–1816
20. Muraoka M, Kelland MA, Yamamoto Y, Tenma N (2020) Tetrahydrofuran hydrate crystal growth inhibitor performance and

mechanism of quaternary ammonium and phosphonium salts. Cryst Growth Des 20(8): 5000–5005

21. Muraoka M, Kelland MA, Yamamoto Y, Suzuki K (2021) Critical growth rate of hydrate crystal growth inhibitors in the low growth rate region. Cryst Growth Des 21(9):4979–4985
22. Anderson R, Mozaffar H, Tohidi B (2011) Development of a crystal growth inhibition based method for the evaluation of kinetic hydrate inhibitors. In Proceedings of the 7th International Conference on Gas Hydrates (pp. 17–21). Edinburgh: Domestic Organizing Committee ICGH-7
23. Ke W, Kelland MA (2016) Kinetic hydrate inhibitor studies for gas hydrate systems: a review of experimental equipment and test methods. Energy Fuel 30(12):10015–10028
24. Lederhos JP, Long JP, Sum A, Christiansen RL, Sloan ED Jr (1996) Effective kinetic inhibitors for natural gas hydrates. Chem Eng Sci 51(8):1221–1229
25. Chua PC, Kelland MA (2012) Tetra(iso-hexyl) ammonium bromide—the most powerful quaternary ammonium-based tetrahydrofuran crystal growth inhibitor and synergist with polyvinylcaprolactam kinetic gas hydrate inhibitor. Energy Fuel 26(2):1160–1168
26. Muraoka M, Susuki N, Yamamoto Y (2016) Evaluation of the performance of kinetic inhibitors for clathrate hydrate using unidirectional growth apparatus. RSC Adv 6(68): 63880–63885
27. Muraoka M, Ohtake M, Yamamoto Y (2019) Kinetic inhibition effect of Type I and III antifreeze proteins on unidirectional tetrahydrofuran hydrate crystal growth. RSC Adv 9(20): 11530–11537

Chapter 12

Divergent Mechanisms of Ice Growth Inhibition by Antifreeze Proteins

Ran Drori and Corey A. Stevens

Abstract

Antifreeze proteins (AFPs) are biomolecules that can bind to ice and hinder its growth, thus holding significant potential for biotechnological and biomedical applications. AFPs are a subset of ice-binding proteins (IBPs) and are found in various organisms across different life kingdoms. This mini-review investigates the underlying mechanisms by which AFPs impede ice growth, emphasizing the disparities between hyperactive and moderate AFPs. Hyperactive AFPs exhibit heightened thermal hysteresis (TH) activity and can bind to both the basal and prism planes of ice crystals, enabling them to endure extremely cold temperatures. In contrast, moderate AFPs predominantly bind to the prism/pyramidal planes and demonstrate lower TH activity. The structural diversity of AFPs and the presence of ordered water molecules on their ice-binding sites (IBS) have been subjects of debate among researchers. Multiple hypotheses have been proposed concerning the significance of ordered water molecules in ice binding. Gaining insights into the binding dynamics and the factors influencing TH activity in AFPs is crucial for the development of efficient synthetic compounds and the establishment of comprehensive models to elucidate ice growth inhibition. Here we emphasize the necessity for further research to unravel the mechanisms of AFPs and presents a pathway for constructing models capable of comprehensively explaining their inhibitory effects on ice growth.

Key words Antifreeze proteins, Adsorption rate, Thermal hysteresis, Hyperactive, Moderate, Synergy

1 Inhibition of Ice Growth by Additives

Despite the abundance of ice on Earth, only a handful of natural and synthetic molecules can adsorb to its crystal surface and inhibit its growth. In fact, most soluble molecules are excluded from the growing ice front. Biomolecules that bind and effectively inhibit ice growth are known as antifreeze proteins (AFPs). AFPs are a subset of proteins that bind to ice, termed ice-binding proteins (IBPs). AFPs are found in all kingdoms of life, including fish, insects, and plants, as an adaption to endure subfreezing temperatures [1, 2]. In these organisms and environments, the growth of small seed ice crystals is inhibited by AFPs, which irreversibly bind to the ice

Ran Drori and Corey A. Stevens (eds.), *Ice Binding Proteins: Methods and Protocols*, Methods in Molecular Biology, vol. 2730,
https://doi.org/10.1007/978-1-0716-3503-2_12,

surface [3–5]. It is thought that AFPs adsorb to an ice and force water molecules to add to the ice surface only between adsorbed AFP molecules [6]. According to the prevalent adsorption–inhibition model [7], the addition of water molecules between ice-bound AFPs leads to a micro-curvature of the ice front. At a certain point, the addition of water molecules to the curved ice surface becomes thermodynamically unfavorable (the Kelvin effect), and ice growth is halted. Ice growth will remain arrested until sufficient supercooling is achieved to engulf the AFP molecules leading to rapid ice growth. As a result of AFP adsorption, the non-equilibrium freezing point is depressed. The gap between the equilibrium melting point and the non-equilibrium freezing point is referred to as thermal hysteresis (TH), during which the ice crystal is neither melting nor growing [8].

Generally, AFPs are classified into two groups based on their activity: hyperactive and moderately active (often called moderate AFP). Hyperactive AFPs (mainly from insects) are roughly ~10 times more potent in TH activity than moderate AFPs (from fish and plants) [8, 9] at equimolar concentrations. The enhanced TH activity of hyperactive AFPs might stem from their ability to bind to both the basal and prism planes of an ice crystal [9, 10], allowing the organism to endure temperatures below −20 °C [11]. However, there are exceptions, as some AFPs bind to the basal plane but demonstrate a low TH activity, such as the AFP found in the Arctic fungus *Psychromyces glacialis* [12, 13].

Moderate AFPs typically bind to prism/pyramidal planes of an ice crystal [4, 14, 15], causing smaller but sufficient freezing point depression of ~1 °C [16]. In addition to having varying levels of TH activity, AFPs have been shown to have extraordinary structural diversity (*see* Fig. 1). The remarkable structural variety is puzzling, and a common structural principle, motif, or molecular mechanism required for binding has yet to be determined. There is an ongoing debate in the AFP field concerning whether ordered water molecules on the ice-binding site (IBS) of AFPs are required for ice binding. The Davies group postulated the "anchored clathrate water" hypothesis after observing these waters in crystal structures of AFPs [17, 18]. This hypothesis suggests waters are ordered on the IBS of an AFP into an "ice-like" lattice that merges with the water molecules in the quasi-liquid layer on the ice surface, which allows for ice binding. Using SFG (sum frequency generation) spectroscopy, Meister et al. found an ice-like layer on the IBS of AFPIII at room temperature [19] and on the surface of a hyperactive AFP bound to ice [20]. NMR measurements on frozen solutions containing AFPIII further supported Meister's finding of an ice-like layer on the IBS [21]. The Tsuda group found ordered water molecules on the IBS of several AFPs and used point mutations on the IBS to demonstrate that AFP activity is eliminated when the organization of water molecules is interrupted

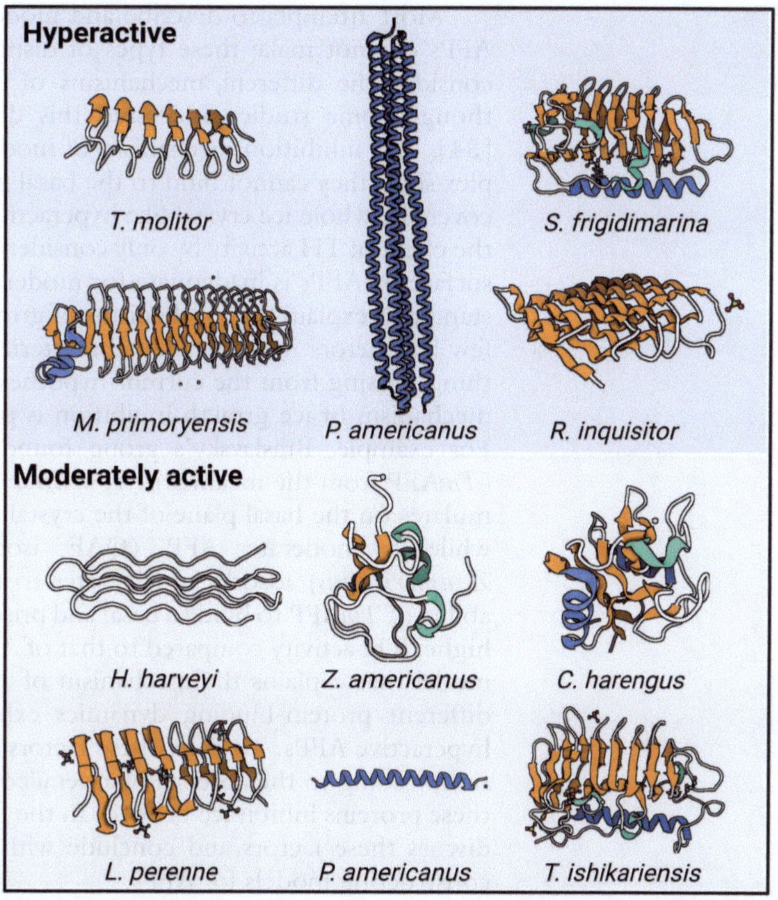

Fig. 1 AFP structures of hyperactive (top) and moderately active (bottom)

[22, 23]. In contrast, using molecular dynamics simulations, the Molinaro group found that ordered waters on the AFP ice-binding site are not required for ice binding [24] for some AFPs, while for others, such ordering contributes to ice binding [25]. The possibility that some AFPs order water molecules before binding to ice while others do not is reasonable and needs to be clarified [26]. Efforts centered on designing efficient synthetic AFP-like compounds have achieved limited efficacy of inhibition [27, 28]. These synthetic molecules have significant abilities to limit ice recrystallization but do not have TH activity. Single-molecule experiments with AFPs suggested that compounds that reversibly bind to ice can inhibit ice recrystallization, but do not exhibit TH activity. Thus, irreversible binding to ice is a requirement for a measurable TH activity [29].

The key to solving this structural puzzle is to understand the mechanisms by which AFPs bind to ice and inhibit its growth. The factors that determine the TH activity of moderate AFPs are discussed below.

Most attempts to describe and model the inhibition effect of AFPs did not make these types of distinctions and did not fully consider the different mechanisms of AFPs [25, 30–33], even though some studies addressed this difference to some extent [34]. The inhibition mechanism of moderate AFPs is more complex since they cannot bind to the basal plane [3, 8, 9] and do not cover the whole ice crystal like hyperactive AFPs. Thus, explaining the effect of TH activity by only considering the coverage of the ice surface by AFPs is inadequate for moderate AFPs [35]. To understand and explain the mechanisms of growth inhibition by AFPs, a few key factors must be fully characterized and understood. One thing missing from the current hypotheses to describe the precise mechanism of ice growth inhibition is protein-binding dynamics. For example, Braslavsky's group found that a hyperactive AFP (*Tm*AFP from the mealworm *T. molitor*) slowly adsorbs and accumulates on the basal plane of the crystal during a period of hours, while a moderate AFP (QAE isoform of AFPIII from *Z. americanus*) adsorbs much faster to the prism plane [9]. The ability of *Tm*AFP to bind to basal and prism planes results in a much higher TH activity compared to that of AFPIII (*see* Fig. 2). Thus, a model that explains the mechanism of all AFPs must include the different protein-binding dynamics exhibited by moderate and hyperactive AFPs. To date, these factors have received little attention, leading to the absence of a detailed model that explains how these proteins inhibit ice growth. In the following sections, we will discuss these factors and conclude with a proposed pathway for constructing models for AFPs.

1.1 Adsorption Rates of AFPs to Ice and Crystal Morphology

Moderate AFPs that cannot bind to the basal plane shape the crystal into a hexagonal bipyramid, in which the sharp tips are the basal surfaces. These AFPs exhibit relatively low levels of TH activity compared to hyperactive AFPs, mainly from insects (*see* Fig. 2). The bipyramidal shape is achieved by the binding of AFPs to non-basal planes of the crystal and inhibiting ice growth along the *a*-axis [9, 14, 36]. As a result, ice growth occurs along the *c*-axis, and each new layer is smaller than the preceding layer as a result of AFP binding to the prism face step. Ice growth continues until the pyramidal planes converge into a sharp tip (*see* Fig. 3) reducing the size of the basal face. The energetic barrier for nucleation of new steps on the basal faces of the newly formed sharp tips increases, and growth along the *c*-axis halts [14, 36]. The formation of a bipyramidal shape by minimizing the area of the basal plane at higher temperatures should provide better growth inhibition, since moderate AFPs cannot bind to the basal surface. In other words, the strategy of these AFPs is "if you cannot bind to it, minimize it." Next, we discuss the effect of crystal morphology and adsorption rate on growth inhibition.

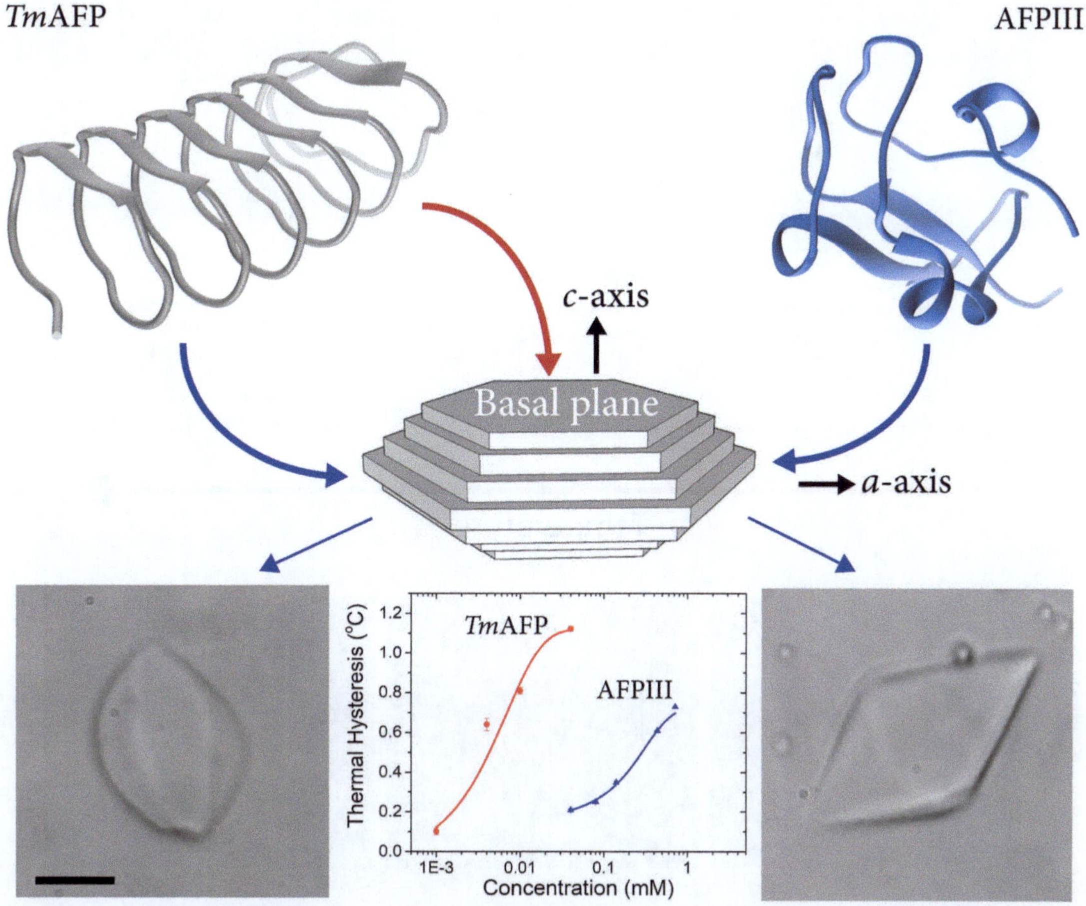

Fig. 2 AFPs bind different crystal planes resulting in different crystal morphologies and TH activity. The hyperactive *Tm*AFP (left) binds to both the basal and prism planes, while the moderate AFPIII (right) only binds to prism/pyramidal planes. The ability of *Tm*AFP to bind to the basal plane leads to increased thermal hysteresis activity compared to AFPIII. Scale bar = 10 μm

Crystal morphology changes by impurities are common, especially the tapering of KDP (potassium dihydrogen phosphate) [37, 38] and quartz crystals [39]. Thus, crystal morphology is an important factor for successful inhibition by moderate AFPs. The quantification of crystal morphology is accomplished by measuring the length (tip to tip, along the *c*-axis) and width (along the *a*-axis) of the bipyramidal crystal. A ratio between these two dimensions was termed the *c-to-a* ratio [40–42] by the Davies group and is used to quantitatively compare crystals shaped by different AFPs. They found that both AFP concentration and mutations to amino acids found on the IBS directly affect crystal morphology. Highlighting, both protein concentration and an appropriate IBS are major factors affecting the shape of ice crystals. Further, AFP solution concentration impacts adsorption rate, suggesting it is vital to measure

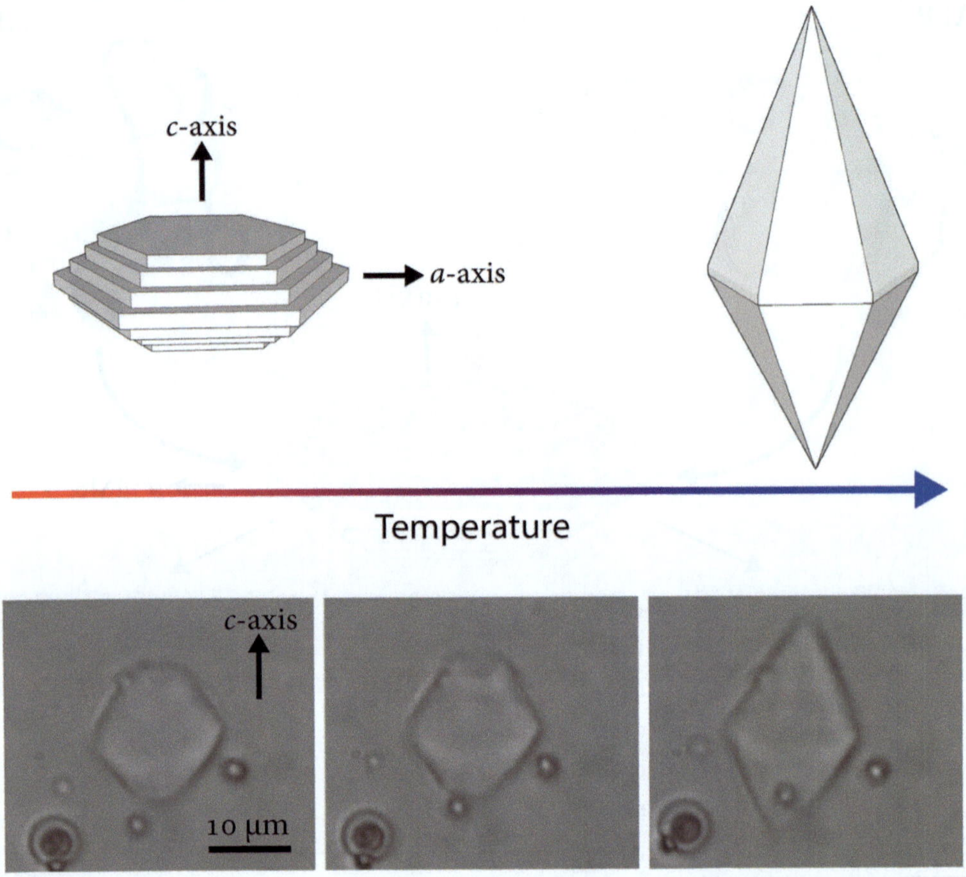

Fig. 3 Moderate AFPs adsorb to prism/pyramidal planes of the crystal, and as the temperature is decreased, ice growth occurs only along the *c*-axis, resulting in a bipyramidal-shaped ice crystal (right)

the intrinsic adsorption constants for each AFP. Figure 4 exhibits the correlation between the adsorption rate, $k_{on}c$, and the quantitative measurement for crystal morphology, *c-to-a* ratio.

Intuitively, AFPs must adsorb to the ice surface at a rate fast enough to slow down the velocity by which the ice front advances (growth along the *a*-axis), as was demonstrated by Sander and Tkachenko [33]. A direct measurement of AFP adsorption rate constants (k_{on}) using fluorescence microscopy was developed by Braslavsky and Drori, leading to the discovery of an experimental link between the adsorption rate ($k_{on}c$) and TH activity of moderate AFPs [3, 9]. The following analysis is focused only on moderate AFPs that cannot bind to the basal plane. A description of the kinetics of AFPs was given by Kubota and others [3, 9, 31] and will be used here:

$$\frac{d\theta}{dt} = k_{\text{on}}(1-\theta)c - k_{\text{off}}\theta$$

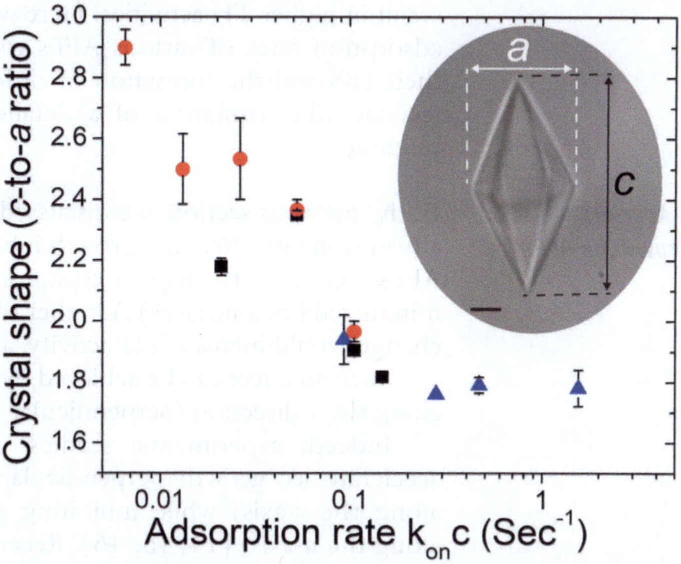

Fig. 4 The effect of adsorption rate on crystal morphology was measured for $AFGP_{1-5}$ (blue, Ref. [49]) and AFPIII-QAE (red, Ref. [41], and black, Ref. [49]). A typical c-to-a measurement is presented in the insert. Scale bar = 10 μm

where θ is the ice surface coverage by AFPs, k_{on} and k_{off} are the intrinsic rate constants of binding, and c is the solution concentration of the AFPs. As irreversible binding to ice was demonstrated for various types of AFPs [3, 5, 29, 36, 43], k_{off} should be negligible for this discussion. The adsorption rate of AFPs to ice is determined by their structure (which dictates the rate constant, k_{on}) and the solution concentration, c. However, the reader should note that the ice surface is dynamic and continuously grows before a complete bipyramidal shape is obtained (*see* Fig. 3). Thus, while AFPs adsorb to the surface and θ is increased, a new ice surface is formed, and the value of θ is decreased. In other words, the value of θ is increasing by the adsorption of more AFPs and simultaneously decreasing by the formation of "fresh" surfaces during the growth along the c-axis. This notion means that surface coverage is not a reliable factor to predict TH activity of moderate AFPs. Indeed, the poor correlation of surface coverage of the moderately active AFPIII and TH activity was demonstrated experimentally [35]. By contrast, a good correlation between TH activity and surface coverage was obtained for the hyperactive *Tm*AFP [35].

In a different experiment that revealed the importance of adsorption rates, borate (0.3 M) was added to an AFGP solution, which lowered the adsorption rate of $AFGP_{1-5}$ by 50% [3]. Borate is known to form bonds with vicinal hydroxyl groups such as those found at the disaccharide moiety in AFGPs, and previous measurements of AFGP solutions with borate exhibited a dramatic decrease in TH activity [44]. Thus, faster adsorption rates of AFPs to ice will result in higher TH activities. Here we propose that comparing the adsorption rates of various AFPs to the structural differences of their IBS and the formation of ordered waters on their IBS will facilitate the formation of a detailed mechanistic model of ice binding.

1.2 Acceleration of Ice Growth by AFPs

In the previous section, we discussed how crystal morphology and adsorption rate affect the growth inhibition by AFPs. Potentially, if AFPs were able to shape a crystal into a bipyramid (and thereby minimized basal surfaces) at higher temperatures, that morphology change would increase TH activity, as discussed above.

Such an effect can be achieved by the acceleration of ice growth along the *c*-direction (perpendicular to the basal plane).

Indeed, experimental studies suggest that moderate AFPs accelerate ice growth perpendicular to the basal plane (growth along the *c*-axis) while inhibiting prism plane growth (growth along the *a*-axis) [14, 45, 46]. Recent molecular dynamics studies found that hyperactive AFPs also promote ice growth before binding to the surface, attributing ice promotion to the formation of the clathrate waters on the IBS [47, 48]. However, the experimental studies mainly measured ice growth velocity at temperatures below the non-equilibrium freezing point (outside the hysteresis gap), at which the mechanism of inhibition by AFPs is different than at temperatures within the hysteresis gap. To understand if growth acceleration is correlated with crystal morphology and subsequently with TH activity, growth velocity measurements should be performed at temperatures higher than the freezing point and should include a broad variety of AFPs. Deng et al. found evidence for growth acceleration by various AFPs at temperatures between the freezing and melting point (i.e., in the hysteresis gap) [49]. The ice growth velocity was increased with AFP concentration for AFPI and AFPIII-QAE. However, growth acceleration was not sensitive to the concentration of the two AFGP isoforms ($AFGP_{2,3,4}$ and $AFGP_8$) [49]. While it is remarkable that AFPs can both inhibit and accelerate ice growth at the same time, this phenomenon is not unique to AFPs and has been observed in calcium oxalate monohydrate (COM) crystals [50]. However, the mechanism by which COM growth inhibition occurs involves aggregation of the molecular inhibitors, whereas AFPs do not tend to aggregate in solution [51]. Furukawa et al. proposed a mechanism of ice growth acceleration by AFPs along the *c*-axis, in which AFPs bind to steps that grow along the *a*-axis and inhibit further growth of the prism face. By interacting and binding to steps, the AFPs also interact with the terrace beneath the growing step (which is a basal face) and act as step nucleation promoters [46].

1.3 Synergy Effect Enhances the Activity of AFPs

In some organisms, multiple isoforms of AFPs have been found. In some fish species producing multiple AFP isoforms, higher amounts of a less TH active (passive isoforms) are expressed as compared to the more TH active isoform, which seems counterintuitive [52, 53]. For instance, in fish containing glycoproteins, the passive $AFGP_8$ is the most prominent isoform detected, with only a small percentage of $AFGP_{1-5}$ present [53]. Similarly, in fish expressing type III AFP, the passive form, SP, is more abundant than the active form, QAE [54]. One hypothesis put forth by the Tsuda group, and others, to explain the difference in abundance of proteins with limited activity is a synergy effect between the active and passive isoforms [52, 54]. TH activity measurements of mixed active and passive isoforms demonstrated that even a small amount of the active form $AFGP_{1-5}$ present led to between two- and eightfold increases in TH activity [52]. Similar experiments conducted by the Tsuda group on the active (QAE) and passive (SP) isoforms of type III AFP showed a more pronounced synergy effect. A minute addition (~1%) of the active isoform to the passive isoform resulted in TH activities similar to those of the pure active isoform [54], and ice growth rate was comparable to that of pure QAE at equivalent concentrations [55]. In an elegant set of experiments done by the Tsuda and Davies groups, a passive isoform was engineered into an active isoform by mutating only four residues on the surface of the protein, and ice growth velocities along the *a*-axis were significantly slowed by mutation of just one residue [56]. Using ice hemispheres and fluorescently labeled AFPs, the Davies group has shown that the QAE isoform binds to both prism and pyramidal planes of the crystal, while the SP isoform primarily binds to pyramidal planes [55, 56]. Wang and Duman mixed four hyperactive AFP isoforms from the beetle *Dendroides canadensis* (DAFP) and observed a synergistic TH enhancement [57]. Using a yeast-two hybrid system, evidence for interaction between isoforms in solution was obtained. Synergistic effects of crystal growth inhibitors were also reported for gas hydrate crystals [58] and calcium oxalate monohydrate (COM) crystals [59]. In all cases, the mechanism of synergy is not fully understood. However, the greatest synergistic effect measured for COM growth inhibition was obtained for compounds that bind to complementary faces of the crystal [59]. The Drori group found that complementary ice binding to multiple crystal planes is a key element of the synergy effect observed in fish AFPs [51]. By fluorescently labeling two AFPIII isoforms, SP and QAE, with different fluorescence dyes, it was demonstrated that each isoform binds to different plane of the ice crystal. This study also demonstrated that the tertiary protein structure of AFPs does not directly affect synergy, and no protein–protein interactions were observed in DLS (dynamic light scattering) experiments [51]. Thus, TH activity enhancement is probably driven by fast adsorption of the active isoform to the prism planes

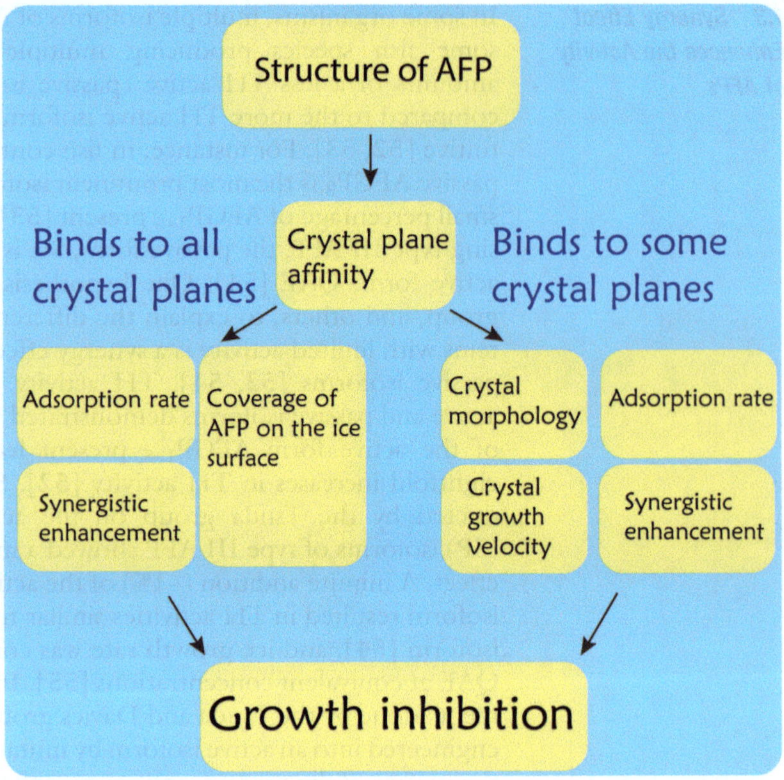

Fig. 5 Flowchart presenting the hypotheses upon which future models to describe the molecular mechanism of ice binding should consider. The small yellow rectangles include factors that ultimately determine the growth inhibition efficacy of the agent

and subsequent adsorption of the passive isoform to pyramidal planes. The passive isoform cannot bind and inhibit the rapidly growing prism faces of the crystal, and no TH is obtained as a result.

1.4 Model for Ice Growth Inhibition by AFPs

Previous models that have tried to predict the activity of AFPs [30, 31, 60, 61] have focused on the attachment of the AFP to ice and neglected to consider the effects mentioned here including adsorption rate [3, 51], crystal morphology [14, 49, 62], growth velocity [45, 46, 49], synergy [51, 54, 57, 62], and the clathrate waters [17–19, 22] on the AFP surface. Thus, a model that predicts ice growth inhibition must include the abovementioned elements and recognize the different inhibition mechanisms by the various AFPs. Figure 5 presents a schematic flowchart illustrating these hypotheses, where each arrow indicates an effect on the downstream box. For example, the structure of the AFP determines its ability to order water molecules on its ice-binding site, and the newly created ice-like layer on the AFP's surface determines to which crystal plane it can bind (crystal plane affinity) and how fast it binds (adsorption rate). To accurately describe the molecular mechanism of ice binding, we think a more holistic model, which considers the multitude of factors described above, is needed. A model of this nature will be inherently interesting from a scientific standpoint but also be useful as a guide for engineering materials to interact with and prevent the growth of ice.

References

1. Bar Dolev M, Braslavsky I, Davies PL (2016) Ice-binding proteins and their function. Annu Rev Biochem 85:515–542
2. Davies PL (2014) Ice-binding proteins: a remarkable diversity of structures for stopping and starting ice growth. Trends Biochem Sci 39:548–555
3. Meister K, DeVries AL, Bakker HJ et al (2018) Antifreeze glycoproteins bind irreversibly to ice. J Am Chem Soc 140:jacs.8b04966
4. Knight CA, Cheng CC, DeVries AL (1991) Adsorption of alpha-helical antifreeze peptides on specific ice crystal surface planes. Biophys J 59:409–418
5. Celik Y, Drori R, Pertaya-Braun N et al (2013) Microfluidic experiments reveal that antifreeze proteins bound to ice crystals suffice to prevent their growth. Proc Natl Acad Sci USA 110: 1309–1314
6. Knight CA, Devries AL (1994) Effects of a polymeric, nonequilibrium antifreeze upon ice growth from water. J Cryst Growth 143:301–310
7. Raymond JA, DeVries AL (1977) Adsorption inhibition as a mechanism of freezing resistance in polar fishes. Proc Natl Acad Sci USA 74: 2589–2593
8. Scotter AJ, Marshall CB, Graham LA et al (2006) The basis for hyperactivity of antifreeze proteins. Cryobiology 53:229–239
9. Drori R, Celik Y, Davies PL et al (2014) Ice-binding proteins that accumulate on different ice crystal planes produce distinct thermal hysteresis dynamics. J R Soc Interface 11: 20140526
10. Basu K, Garnham CP, Nishimiya Y et al (2014) Determining the Ice-binding Planes of Antifreeze Proteins by Fluorescence-based Ice Plane Affinity. Jove J Vis Exp
11. Duman JG (2015) Animal ice-binding (antifreeze) proteins and glycolipids: an overview with emphasis on physiological function. J Exp Biol 218:1846–1855
12. Kaleda A, Haleva L, Sarusi G et al (2018) Saturn-shaped ice burst pattern and fast basal binding of an ice-binding protein from an antarctic bacterial consortium. Langmuir 35: 7337–7346
13. Arai T, Yamauchi A, Yang Y et al (2022) Adsorption of ice-binding proteins onto whole ice crystal surfaces does not necessarily confer a high thermal hysteresis activity. Sci Rep 12:15443
14. Knight CA, DeVries AL (2009) Ice growth in supercooled solutions of a biological "antifreeze", AFGP 1-5: an explanation in terms of adsorption rate for the concentration dependence of the freezing point. Phys Chem Chem Phys 11:5749–5761
15. Rahman AT, Arai T, Yamauchi A et al (2019) Ice recrystallization is strongly inhibited when antifreeze proteins bind to multiple ice planes. Sci Rep 9:2212
16. Knight CA, DeVries AL, Oolman LD (1984) Fish antifreeze protein and the freezing and recrystallization of ice. Nature 308:295–296
17. Garnham CP, Campbell RL, Davies PL (2011) Anchored clathrate waters bind antifreeze proteins to ice. Proc Natl Acad Sci USA 108: 7363–7367
18. Sun TJ, Lin FH, Campbell RL et al (2014) An antifreeze protein folds with an interior network of more than 400 semi-clathrate waters. Science 343:795–798
19. Meister K, Strazdaite S, DeVries AL et al (2014) Observation of ice-like water layers at an aqueous protein surface. Proc Natl Acad Sci 111:201414188
20. Meister K, Moll CJ, Chakraborty S et al (2019) Molecular structure of a hyperactive antifreeze protein adsorbed to ice. J Chem Phys 150: 131101
21. Siemer AB, Huang KY, McDermott AE (2010) Protein-ice interaction of an antifreeze protein observed with solid-state NMR. Proc Natl Acad Sci USA 107:17580–17585

22. Mahatabuddin S, Fukami D, Arai T et al (2018) Polypentagonal ice-like water networks emerge solely in an activity-improved variant of ice-binding protein. Proc Natl Acad Sci 115: 5456–5461
23. Arai T, Nishimiya Y, Ohyama Y et al (2019) Calcium-binding generates the semi-clathrate waters on a type II antifreeze protein to adsorb onto an ice crystal surface. Biomol Ther 9:162
24. Hudait A, Moberg DR, Qiu Y et al (2018) Preordering of water is not needed for ice recognition by hyperactive antifreeze proteins. Proc Natl Acad Sci 115:8266–8271
25. Hudait A, Odendahl N, Qiu Y et al (2018) Ice-nucleating and antifreeze proteins recognize ice through a diversity of anchored clathrate and ice-like motifs. J Am Chem Soc 140: 4905–4912
26. Marks SM, Patel AJ (2018) Antifreeze protein hydration waters: unstructured unless bound to ice. Proc Natl Acad Sci USA 115:8244–8246
27. Congdon T, Notman R, Gibson MI (2013) Antifreeze (glyco)protein mimetic behavior of poly(vinyl alcohol): detailed structure ice recrystallization inhibition activity study. Biomacromolecules 14:1578–1586
28. Biggs CI, Stubbs C, Graham B et al (2019) Mimicking the ice recrystallization activity of biological antifreezes. When is a new polymer "active"? Macromol Biosci 19:1–9
29. Tas RP, Hendrix Marco MR, Voets IK (2023) Nanoscopy of single antifreeze proteins reveals that reversible ice binding is sufficient for ice recrystallization inhibition but not thermal hysteresis. Proc Natl Acad Sci USA 120:1–8
30. Kozuch DJ, Stillinger FH, Debenedetti PG (2018) Combined molecular dynamics and neural network method for predicting protein antifreeze activity. Proc Natl Acad Sci USA 115:13252–13257
31. Kubota N (2011) Effects of cooling rate, annealing time and biological antifreeze concentration on thermal hysteresis reading. Cryobiology 63:198–209
32. Mochizuki K, Molinero V (2018) Antifreeze glycoproteins bind reversibly to ice via hydrophobic groups. J Am Chem Soc 140(4): 4803–4811
33. Sander LM, Tkachenko AV (2004) Kinetic pinning and biological antifreezes. Phys Rev Lett 93:128102
34. Chasnitsky M, Braslavsky I (2019) Ice-binding proteins and the applicability and limitations of the kinetic pinning model. Philos Trans R Soc A Math Phys Eng Sci 377:20180391
35. Drori R, Davies PL, Braslavsky I (2014) Experimental correlation between thermal hysteresis activity and the distance between antifreeze proteins on an ice surface. RSC Adv 5:7848–7853
36. Drori R, Davies PL, Braslavsky I (2015) When are antifreeze proteins in solution essential for ice growth inhibition? Langmuir 31:5805–5811
37. Sangwal K, Torrent-Burgues J, Gorostiza P et al (1999) AFM study of the behaviour of growth steps on the (100) faces of KDP crystals and the tapering phenomenon. Cryst Res Technol 34:667–675
38. Dam B, Bennema P, Van Enckevort WJP (1986) The mechanism of tapering on KDP-type crystals. J Cryst Growth 74:118–128
39. Lu T, Yallee RB, Ong CK et al (1995) Formation mechanism of tapering of crystals: a comparative study between potassium dihydrogen phosphate crystals and natural quartz crystals. J Cryst Growth 151:342–347
40. Heman C, DeLuca CI, Davies PL (1995) Mixing antifreeze protein types changes ice crystal morphology without affecting antifreeze activity. FEBS Lett 357:183–186
41. DeLuca CI, Chao H, Sönnichsen FD et al (1996) Effect of type III antifreeze protein dilution and mutation on the growth inhibition of ice. Biophys J 71:2346–2355
42. Bar-Dolev M, Celik Y, Wettlaufer JS et al (2012) New insights into ice growth and melting modifications by antifreeze proteins. J R Soc Interface 9:3249–3259
43. Pertaya N, Marshall CB, DiPrinzio CL et al (2007) Fluorescence microscopy evidence for quasi-permanent attachment of antifreeze proteins to ice surfaces. Biophys J 92:3663–3673
44. Ebbinghaus S, Meister K, Born B et al (2010) Antifreeze glycoprotein activity correlates with long-range protein-water dynamics. J Am Chem Soc 132:12210–12211
45. Vorontsov DA, Sazaki G, Titaeva EK et al (2018) Growth of ice crystals in the presence of type III antifreeze protein. Cryst Growth Des 18:2563–2571
46. Furukawa Y, Nagashima K, Nakatsubo SI et al (2017) Oscillations and accelerations of ice crystal growth rates in microgravity in presence of antifreeze glycoprotein impurity in supercooled water. Sci Rep 7:43157
47. Cui S, Zhang W, Shao X et al (2022) Hyperactive antifreeze proteins promote ice growth before binding to it. J Chem Inf Model 62: 5165–5174

48. Cui S, Zhang W, Shao X et al (2022) Do antifreeze proteins generally possess the potential to promote ice growth? Phys Chem Chem Phys 24:7901–7908
49. Deng J, Apfelbaum E, Drori R (2020) Ice growth acceleration by antifreeze proteins leads to higher thermal hysteresis activity. J Phys Chem B 124:11081–11088
50. Weaver ML, Qiu SR, Hoyer JR et al (2009) Surface aggregation of urinary proteins and aspartic acid-rich peptides on the faces of calcium oxalate monohydrate investigated by in situ force microscopy. Calcif Tissue Int 84: 462–473
51. Berger T, Meister K, DeVries AL et al (2019) Synergy between antifreeze proteins is driven by complementary ice-binding. J Am Chem Soc 141:19144–19150
52. Osuga DT, Ward FC, Yeh Y et al (1978) Cooperative functioning between antifreeze glycoproteins. J Biol Chem 253:6669–6672
53. DeVries AL, Komatsu SK, Feeney RE (1970) Chemical and physical properties of freezing point-depressing glycoproteins from Antarctic fishes. J Biol Chem 245:2901–2908
54. Takamichi M, Nishimiya Y, Miura A et al (2009) Fully active QAE isoform confers thermal hysteresis activity on a defective SP isoform of type III antifreeze protein. FEBS J 276: 1471–1479
55. Garnham CP, Natarajan A, Middleton AJ et al (2010) Compound ice-binding site of an antifreeze protein revealed by mutagenesis and fluorescent tagging. Biochemistry 49:9063–9071
56. Garnham CP, Nishimiya Y, Tsuda S et al (2012) Engineering a naturally inactive isoform of type III antifreeze protein into one that can stop the growth of ice. FEBS Lett 586:3876–3881
57. Wang L, Duman JG (2005) Antifreeze proteins of the beetle dendroides canadensis enhance one another's activities. Biochemistry 44: 10305–10312
58. Chua PC, Kelland MA (2012) Tetra(iso-hexyl) ammonium bromide – the most powerful quaternary ammonium-based tetrahydrofuran crystal growth inhibitor and synergist with polyvinylcaprolactam kinetic gas hydrate inhibitor. Energy Fuel 26:1160–1168
59. Farmanesh S, Ramamoorthy S, Chung J et al (2014) Specificity of growth inhibitors and their cooperative effects in calcium oxalate monohydrate crystallization. J Am Chem Soc 136:367–376
60. Liu J, Li Q (2006) Theoretical model of antifreeze protein-ice adsorption: binding of large ligands to a two-dimensional homogeneous lattice. Chem Phys Lett 422:67–71
61. Lopez Ortiz JI, Quiroga E, Narambuena CF et al (2021) Thermal hysteresis activity of antifreeze proteins: a model based on fractional statistics theory of adsorption. Phys A Stat Mech Appl 575:126046
62. Chao H, DeLuca CI, Davies PL (1995) Mixing antifreeze protein types changes ice crystal morphology without affecting antifreeze activity. FEBS Lett 357:183–186

Part III

Chemical Modifications and Synthesis of IBP Mimics, MD Simulations and Evolution of IBPs

Chapter 13

Multiscale Molecular Dynamics Simulations of Ice-Binding Proteins

Arpa Hudait

Abstract

Ice-binding proteins (IBPs) are a diverse class of proteins that are essential for the survival of organisms in cold conditions. IBPs are diverse in their function and can prevent or promote ice growth and selectively bind to specific crystallographic planes of the growing ice lattice. Moreover, IBPs are widely utilized to modulate ice crystal growth and recrystallization in the food industry and as cryoprotectants to preserve biological matter. A key unresolved aspect of the mode of action is how the ice-binding sites of these proteins distinguish between ice and water and interact with multiple crystal facets of the ice. The use of molecular dynamics (MD) simulation allows us to thoroughly investigate the binding mechanism and energetics of ice-binding proteins, to complement and expand on the mechanistic understandings gained from experiments. In this chapter, we describe a series of molecular dynamics simulation methodologies to investigate the mechanism of action of ice-binding proteins. Specifically, we provide detailed instructions to set up MD simulations to study the binding and interaction of ice-binding proteins using atomistic and coarse-grained simulations.

Key words Ice-binding proteins, Molecular dynamics simulation, Interfacial water

1 Introduction

Molecular dynamics (MD) simulation, a technique first developed in the 1970s, is used to mimic the interaction of atoms in real life [1, 2]. In MD simulations, a collection of atoms or molecules are deterministically evolved in time using numerical approximations of Newton's equation of motion. This technique facilitates the study of molecular processes in multiple time and length scales. The user can investigate the vibration and rotation of atoms in a small organic molecule, estimate thermodynamic properties of molecular liquids, and explore conformational behavior of a solvated protein. The user can also simulate large-scale collective phenomena such as crystallization of materials, protein-induced membrane remodeling, and the self-assembly behavior of proteins. Here, we briefly elaborate on the theoretical details of the MD simulations. The

Ran Drori and Corey A. Stevens (eds.), *Ice Binding Proteins: Methods and Protocols*, Methods in Molecular Biology, vol. 2730, https://doi.org/10.1007/978-1-0716-3503-2_13,

main goal of MD simulations is to find the position of the atoms of the system as a function of time and generate a trajectory at discrete time intervals. We first define a potential energy function to define the bonded and nonbonded interactions between the atoms. This potential energy function is also known as the force field (FF). From the initial coordinate and velocity (typically assigned based on the temperature of the simulation), we first calculate the force on each atom, based on the user-assigned force field. Then the position of each atom at the next time step is updated from the force and initial velocity of the system. The force on each atom at the next time step is then calculated from the updated atom coordinates. Finally, the velocity on each atom at the next time step is updated using the force from the current and previous time steps. MD simulations can be performed for a particular system for a few picoseconds to multiple milliseconds depending on the degree of complexity of the force field, system size, and computing resources. Furthermore, in MD simulations, the user can choose to maintain the temperature, pressure, or volume at a specific value using appropriate methods. Overall, MD simulations have increasingly been utilized to gain insight on processes that are challenging to study through state-of-the-art experimental techniques and to motivate future experimental directions.

Our particular research interests are to study the mechanism of action of ice-binding proteins (IBPs). Antifreeze proteins (AFPs) are a class of IBPs that inhibit ice growth by irreversibly binding to the surface of a growing ice crystal and compelling the crystal to grow with curvature and inhibiting ice growth, thereby ensuring the survival of the organism. Since the first discovery of AFPs in the Antarctic notothenioid fish in the 1960s [3, 4], AFPs have been identified in wide-ranging organisms such as insects, bacteria, and fish [5, 6]. This demonstrates that many organisms living in cold environments have developed these proteins as a survival strategy. Another class of IBPs is the ice-nucleating proteins (INPs) that have evolved to promote the nucleation and growth of ice. *Pseudomonas syringae* (*Ps*INP) and *Pseudomonas borealis* (*Pb*INP) are the most efficient ice-nucleating agents, nucleating ice at −2 °C [7, 8]. The AFPs and INPs are functionally contrasting, yet both classes of proteins can share a common ice-binding site. Hyperactive AFPs from the insect and bacterial INPs bind to ice through parallel stacks of β-helices that contain TxT repeats (T = threonine and x is a nonconserved amino acid). The TxT repeats create a lattice-like arrangement of the hydroxyl groups from the threonine residues with similar lattice matching to the basal and prismatic plane of the ice. This allows for seamless integration and binding of the IBP to ice [9–11].

Previously, it was proposed that TxT motifs bind to ice through an anchored clathrate (AC) motif, based on the arrangement of crystallographic water at the ice-binding surface of the bacterial

AFP *Marinomonas primoryensis* (*Mp*AFP) [12]. In the AC motif, the clathrate-like arrangement of water around the methyl group is anchored to the protein hydroxyl group through hydrogen bonds. It is noteworthy that the clathrate-like arrangement around the methyl groups is driven by hydrophobic interactions, while hydrophilicity of the hydroxyl group is necessary for the hydrogen bonding interactions. The molecular mechanism of ice recognition and binding by IBPs is dependent on the arrangement of amino acids, water dynamics, and relative number of hydrophilic or hydrophobic residues at the binding site. Finally, a detailed understanding of the relationship between the chemical nature of the ice-binding sites of diverse IBPs and ice recognition and binding mechanism can also aid the design of novel bioinspired synthetic ice-inhibiting molecules.

In this chapter, we describe in detail how to set up and run all-atom and coarse-grained MD simulations to examine the mode of ice recognition and binding by the hyperactive AFPs, performing binding free calculations of hyperactive AFPs to ice using umbrella sampling MD simulations, and perform ice nucleation simulations induced by the bacterial INPs. Finally, we include notes for the reader about the likely challenges during the system preparation and the simulation runs. We note that these method protocols are based on our recently published articles on hyperactive AFPs and bacterial INPs [13–16].

2 Materials

To prepare the initial system for MD simulation of IBPs, the user will need starting coordinates of the respective proteins. The initial coordinates of the experimentally resolved starting coordinates are publicly available in the Protein Data Bank (PDB, www.rcsb.org). The three hyperactive insect antifreeze proteins used here are *Tenebrio molitor* (*Tm*AFP), *Choristoneura fumiferana* (*Cf*AFP), and *Rhagium inquisitor* (*Ri*AFP). The initial coordinates of *Tm*AFP, *Cf*AFP, and *Ri*AFP are derived from the respective crystal structures IEZG [9], 1M8N [10], and 4DT5 [11]. Figure 1 shows the structure of the hyperactive AFPs used in our studies. The initial model of the bacterial ice-nucleating protein *Ps*INP is built from the structure reported in reference [17]. This model repeats a monomeric unit consisting of a 16-amino acid sequence (GYGSTQTSGSESSLTA, a single β-helical fold), where TQTS and ESSLT are the putative ice-binding sites [18].

To perform all-atom simulations of the *Tm*AFP in solution or in coexistence with an ice slab, we modeled the protein with the CHARMM22 force field with the CMAP correction [19, 20]. A key requirement for ice recognition and binding is the rotameric

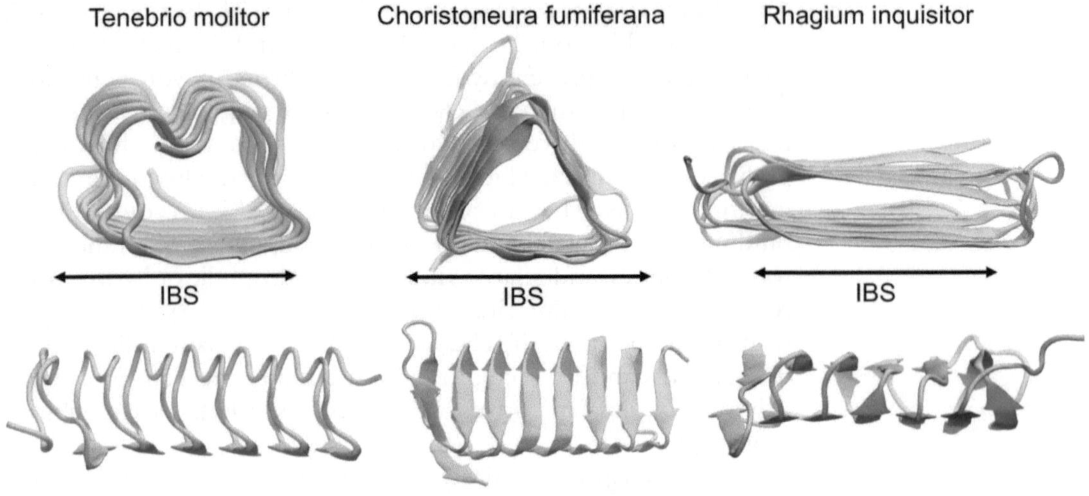

Fig. 1 Structure of the hyperactive proteins *Tenebrio molitor* (TmAFP), Choristoneura fumiferana (CfAFP), and Rhagium inquisitor (RiAFP) showing the front and side view. The ice-binding site (IBS) containing the array of threonine residues in each protein is marked

uniformity of the ice-binding threonine residues [21]. We note that only the CHARMM22 force field maintains the rotameric uniformity of the threonine residues. However, this is not the case when *Tm*AFP is modeled with the CHARMM36 and the AMBER force fields (ff12SB and ff14SB) [22]. The water is modeled at the rigid nonpolarizable level with TIP4P/2005 force field [23] and at the polarizable level with the MB-pol potential [24–26]. The user can also choose to perform the simulations with other rigid nonpolarizable models like TIP4P/Ice, TIP4P/Ew, and SPC/E [27–29].

All-atom simulations reported in this chapter at the rigid nonpolarizable level were performed with the GPU-accelerated AMBER simulation package [30]. However, the user may choose to do the simulations in other popular simulation packages such as GROMACS [31] and NAMD [32]. The all-atom polarizable simulations were performed in the DL_POLY simulation package [33]. The user can also perform the simulations with MB-pol many-body water potential with the LAMMPS simulation package when the MBX library is interfaced with the LAMMPS [34, 35]. In the coarse-grained simulations, water is modeled with mW water model, which is extensively utilized for studying nucleation and growth of ice, the structure of ice polymorphs, and ice–water interfacial properties [36–43]. The protein is modeled as a rigid body at the united atom level, i.e., with all heavy atoms except hydrogen atoms. Finally, the user can use the Visual Molecular Dynamics (VMD) software to visualize the trajectories generated from the simulations and render images for publications [44].

3 Methods

3.1 System Preparation

3.1.1 All-Atom Simulations with Nonpolarizable Water Model

To set up all-atom simulations of protein binding to an ice–water composite system, the user will need to separately equilibrate the ice and water slab before adding the protein at the ice–water interface. To this end, we describe the steps to build the all-atom protein model and the initial system for ice-binding simulation of the protein.

1. In the psfgen plugin of the Visual Molecular Dynamics (VMD) software, input the desired protein coordinate downloaded from the RCSB PDB. This will generate the protein coordinates with the hydrogens.
2. To build the ice/water composite system, the user will need a block of hexagonal ice and a slab of liquid water. We note that the block of hexagonal ice can be generated by replicating a unit cell to the desired dimension. Generate the slab of liquid water by melting the hexagonal ice block at 298 K.
3. Manually combine the block of hexagonal ice and liquid water to generate a composite system, exposing either the basal or the prismatic plane of hexagonal ice. We note that the properties of the ice layer in coexistence with the liquid water are distinct from that of bulk water. Hence, to distinguish between interfacial ice layers and bulk ice, we recommend that the block of ice must contain at least six bilayers (four interfacial ice bilayers in contact with the liquid water and two bilayers representing bulk ice).
4. At the same time, generate a hydrated protein system by placing the protein in a cubic simulation cell of water. This step is typically the first step of generating any biomolecular simulation system and is well documented in the online manual of popular MD simulation packages.
5. Simulate the solvated protein for 5 ns in a cubic simulation cell by harmonically restraining the protein. Extract the protein coordinates with the hydration shell water (defined as all-water molecules within 5 Å of the protein heavy atoms).
6. Place the protein in the liquid phase in such a way that the ice-binding site is oriented toward the plane of the ice/water interface. We recommend that the initial distance between the ice-binding site and the ice/water interface is 7.1 Å (approximately the height of two bilayers of ice). Then remove all the water molecules within 3.5 Å of the region occupied by the protein to avoid system instability due to close repulsive contacts. Figure 2 shows the initial placement of the protein near the ice/water interface.

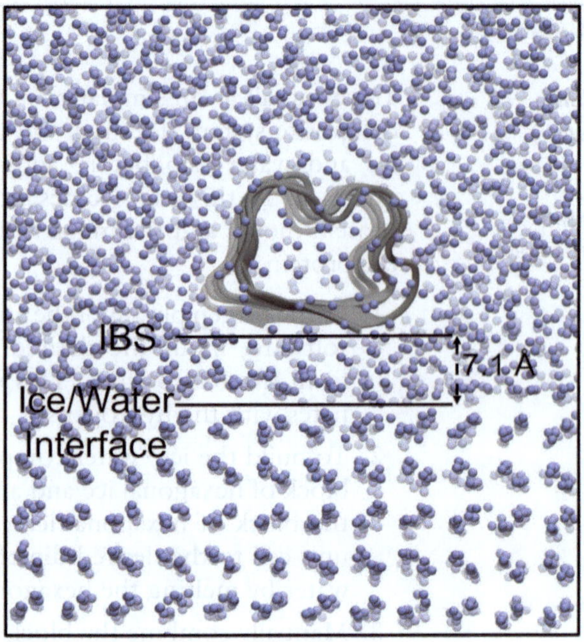

Fig. 2 Initial system of the ice-binding all-atom simulations. The ice/water interface and the IBS of the AFP are marked. In the initial configuration, the IBS of the AFP is 7.1 Å away from the ice/water interface

7. Add sodium or chloride ions in the liquid phase to make the system charge neutral.
8. Minimize the composite system to remove bad protein/water contacts and perform a restrained equilibration (by harmonically restraining the protein) for 200 ps for relaxation of the water molecules. The final configuration of the restrained simulation is the initial configuration for the production runs of the ice-binding simulations.
9. The simulations of ice growth can be performed at the temperature range T_m-2 to T_m-10 K, where T_m is the melting temperature of ice for the water model. The user must keep in mind that at temperatures closer to the melting point, the rate of ice growth is prohibitively slow, and therefore, simulations of protein binding to ice can be computationally expensive. Conversely, at high supercooling, ice will grow rapidly engulfing the protein, thus preventing analysis of the water properties at the ice/water and ice/water/protein interface.
10. Perform the production simulations with the NPT ensemble by applying a periodic boundary condition in all directions. For these simulations, control the system pressure (1 bar) through an isotropic Berendsen barostat with a relaxation time of 1 ps and the system temperature with a Langevin thermostat with a damping frequency of 0.1 ps^{-1} [45–47]. Choose the

simulation time step of 2 fs and integrate Newton's equations of motion with the velocity Verlet algorithm. We recommend using the SHAKE algorithm to constrain bonds involving heavy atoms and hydrogen atoms [48, 49].

11. Use the particle mesh Ewald (PME) option to treat long-range electrostatics [50] and a 9 Å cutoff to treat the van der Waals interaction.

3.1.2 All-Atom Simulations with Polarizable Water Model

We performed many-body molecular dynamics (MB-MD) simulations to compute the infrared and Raman spectra of the water molecules at the ice-binding site and non-ice-binding site of the protein. In the MB-MD simulations, water is modeled with the polarizable MB-pol model. The system setup of these simulations is identical to that of the protein simulations with the rigid nonpolarizable TIP4P/2005 model. However, MB-MD simulations are significantly more computationally expensive than all-atom simulations with TIP4P/2005 model since MB-pol is a polarizable model. Hence, the system size of the MB-MD simulations is approximately a third of that in the all-atom simulations with the TIP4P/2005 model. The steps to generate the initial configuration of the MB-MD simulation are described below.

1. Equilibrate a system of the protein at the ice/water interface for 50 ns at 246 K to allow the ice to grow, resulting in the binding of the protein to the ice.
2. Use the equilibrated configuration as the initial configuration for the MB-MD simulations by replacing the TIP4P/2005 water molecules with the MB-pol water.
3. In the MB-MD simulations, model the protein and ions with the CHARMM22 force field, and use the identical cross interactions between protein and ions to TIP45/2005 water.
4. Prepare the initial configuration of the protein solvated with the MB-pol water by equilibrating the system for 100 ps in the NVT ensemble at 246 K. Note that the temperature is 11.5 K below the melting point of ice I_h for MB-pol water. However, the simulation times are insufficient for the ice to grow.
5. Perform production runs in the NVE ensemble to generate trajectories from which the infrared and Raman spectra are calculated. In the MB-MD simulations, integrate Newton's equations of motion with a time step of 0.2 fs, and use the "time-reversible always stable predictor–corrector" method with fourth-order extrapolation in the predictor S4 to compute the induced dipoles [51]. Maintain the temperature using the Nose–Hoover thermostat, by coupling each degree of freedom to four Nose–Hoover chain thermostats [52, 53].

A sample input script to run MB-pol simulations in DL_POLY is shown below:

```
DLTTM CONTROL
iseed -2647511 ! seed for the random number generator
cmd ! CMD simulation (NVE or NVT if cmdnvt is defined)
temperature 246.00 !Target temperature in K
#rstrt ! if restarting from a previous configuration remove #
# SIMULATION PARAMETERS
steps 500000 ! number of steps
timestep 0.0002 ! time step in ps
# TRAJECTORY PARAMETERS
stats 100 ! printing STATIS.IBEAD
# for CMD:
# (0 ndump 0) prints centroid positions in POSITION_CMD
# (0 ndump 1) prints centroid positions & velocities
# in POSITION_CMD and VELOCITY_CMD
traj 0 10 1
# SYSTEM CUTOFFS
cutoff 9.00 ! electrostatics cutoff
rvdw 9.00 ! nonbonded cutoff
delr width 0.50 ! for Verlet list
# EWALD
ewald precision 1.0E-6 ! dlpoly define alpha & k-vectors based
on the precision
# POLARIZABLE SIMULATIONS.
polarizability ! define a polarizable model
aspc ! always stable predictor-corrector algorithm
vmbpol ! MB-pol (requires polarizability)
# NOSE-HOOVER CHAINS
nchain 4
# CLOSING FILES
job time 1.0D+9
close time 1.0D+2
finish
```

3.1.3 Coarse-Grained Simulations

The coarse-grained simulations are performed with the united-atom protein and single-site mW water. The mW model uses the Stillinger–Weber potential for interatomic interactions [54]. In the SW potential, pairwise interactions are modeled with a short-range two-body (ϕ_2) term and a three-body (ϕ_3) term that penalizes any angular configurations between water beads that deviate from tetrahedrality. The full expression for ϕ_2 and ϕ_3 are as follows:

$$E = \sum_i \sum_{j>i} \phi_2(r_{ij}) + \sum_i \sum_{j \neq i} \sum_{k>j} \phi_3(r_{ij}, r_{ik}, \theta_{ijk})$$

$$\phi_2(r_{ij}) = A\epsilon_{ij}\left[B\left(\frac{\sigma_{ij}}{r_{ij}}\right)^p - \left(\frac{\sigma_{ij}}{r_{ij}}\right)^q\right] exp\left(\frac{\sigma_{ij}}{r_{ij} - a\sigma_{ij}}\right)$$

$$\phi_3(r_{ij}, r_{ik}, \theta_{ijk}) = \lambda_{ijk}\epsilon_{ijk}\left[\cos\theta_{ijk} - \cos\theta_{0ijk}\right]^2 exp\left(\frac{\gamma\sigma_{ij}}{r_{ij} - a\sigma_{ij}}\right) exp\left(\frac{\gamma\sigma_{ik}}{r_{ik} - a\sigma_{ik}}\right)$$

The summations are over all neighbors j and k of the central atom i within the cutoff distance of $a\sigma$. The mW model is over 100 times faster than atomic nonpolarizable models such as TIP4P/2005 and SPC/E due to short-range interactions and coarser molecular representation. In our coarse-grained simulations, the protein–water interactions are also modeled with the SW potential. The strength of the interaction between the hydroxyl groups of the protein (OH) and mW water is chosen to be 10% higher than the interaction between mW water ($\varepsilon_{\text{OH–mW}} = 1.1\varepsilon_{\text{mW–mW}}$) to reflect a stronger hydrogen bond for hydroxyl–water interactions relative to water–water interactions [55]. The terminal methyl group is modeled as M methane, which was parameterized to induce clathrate formation [56]. Taken together, the coarse-grained resolution of the models and short-ranged nature of the interactions allow extensive sampling of the dynamics of ice recognition while using a significantly larger system size.

Simulations of AFP Binding to Ice The steps to prepare the coarse-grained simulations of binding of AFPs to ice are described below:

1. The coarse-grained simulations are performed using the LAMMPS simulation engine [34]. Therefore, the user will first need to download and compile the LAMMPS package in any available high-performance computing (HPC) cluster. The detailed instructions to build LAMMPS are provided here (https://docs.lammps.org/Build.html). We note that during LAMMPS compilation, additional packages can be added. To run the coarse-grained simulations described here, the user will need to add the following packages, MANYBODY, MOLECULE, and RIGID, during compilation.
2. Use the identical steps described previously for all-atom simulations to generate the composite ice/water system, with the protein solvated in the liquid phase. In the coarse-grained simulations, the user may choose to build a significantly large system with more bilayers of ice (double or more than the number of bilayers and surface area relative to the system sizes used for all-atom simulations).

3. Perform the simulations at 272.5 K (0.5 below the melting point of mW model) using the NPT ensemble. Integrate the equations of motions with the velocity Verlet algorithm using a time step of 10 fs. Control the temperature and pressure with the Nose–Hoover thermostat and barostat with damping times of 10 and 50 ps, respectively. The pressures are controlled independently in all three dimensions. Integrate the equations of motion of the rigid body protein as a single molecular entity with translational and rotational motions following the rigid body dynamics algorithm implemented in LAMMPS [57, 58].
4. To run the simulations, the user will need to prepare an input file that provides all the simulation settings and a force field file to write out intermolecular interaction parameters. A sample input file is shown below:

```
units real # sets the units of the output
atom_style full # attributes of the atoms
boundary p p p # periodic b.c. in xy and fixed in z (xy is plane
of the surface)
read_data system data # initial system
pair_style hybrid sw sw0
pair_coeff * * sw0 TmAFP-mW.sw CT CM CO NH NT OC SH OH mW.
pair_coeff * * sw TmAFP-mW.sw NULL NULL NULL NULL NT OC NULL OH
mW # (the ones that do not have three-body interactions are
written as NULL)
neighbor 2.0 bin # defines a tolerance for neighbor buffer
region
neigh_modify every 2 delay 10 check yes # how often builds the
neighbor list
timestep 10
fix 101 all momentum 1000 linear 1 1 1 angular # zeroes the
linear and angular momentum
group protein type 1 2 3 4 5 6 7 8
group water type 9
velocity all create 298 4928459 rot yes mom yes dist gaussian
fix 3 water npt temp 298 298 1000 aniso 1 1 5000
fix 4 protein rigid/nvt single temp 298 298 1000
restart 100000 config # frequency of outputting restart files
with coordinates and velocities
dump 1 all custom 10000 dump.lammpstrj id mol type xs ys zs
#output trajectory
run 1000000
```

Umbrella Sampling Simulations to Compute Free Energy of Binding of AFPs to Ice The umbrella sampling (US) method is a commonly used enhanced sampling simulation technique to estimate the binding free energy [59]. In the umbrella sampling method, a series of discrete time windows are used along a selected collective variable (CV) to sample the process of binding and unbinding. For each window, a harmonic bias potential is defined by an appropriately chosen force constant (K). The umbrella sampling technique can be used to estimate the binding affinity of antifreeze proteins to different planes of ice. The steps to set up US simulations is as follows:

1. Create a simulation cell with several bilayers of ice such that the ice bilayers are in contact with liquid on one side and vapor on other side. In the initial configuration, the protein is placed ~12 Å away from the ice. In the simulation, fix the bottom two layers in contact with vapor.
2. Constrain the amount of ice (except the immobilized bottom two layers) in the simulation cell with a harmonic potential on the global Steinhardt bond-order parameter q_6 of the system [60]. Set the equilibrium reference value of q_6 from the initial configuration (to calculate q_6, see https://docs.lammps.org/compute_orientorder_atom.html). Use a force constant of 20 kcal mol^{-1}. The actual force constant used by the simulation is multiplied by the number of mobile water molecules in the system.
3. Restrain the distance (CV) between the center of mass of the protein and the ice surface with a harmonic spring. Use a force constant of 50 kcal mol^{-1}, varying the equilibrium distance by 0.25 Å (this is the spacing between two consecutive windows). We note that the user may choose to use a lower force constant and higher distance spacing between the windows that is appropriate for the system. For each window, run 25-ns-long simulations at 273 K (melting point of mW water).
4. Compute the free energy profile of protein binding and unbinding from the histogram of the CV at each window using the WHAM algorithm [61]. The weighted histogram analysis method (WHAM) is a post-processing method that is used to calculate the unbiased free energy by the umbrella integration.

   The LAMMPS input file to perform the umbrella sampling simulations is as follows:

```
fix 1 wall setforce 0.0 0.0 0.0. # restrain the bottom two
layers of ice
fix 2 protein rigid/nvt single temp 273 273 500
# harmonic potential to restrain the total amount of ice in the
system
fix 3 water steinhardt/harmonic 6 0.075 142260 3 3.5
fix pull protein spring tether 50.0 NULL NULL ${ref} ${R0}
# ${ref} is the tethering point of the spring at the surface of
ice
```

```
# ${R0} denotes the center of mass of the protein, different
each window
fix 4 water nvt temp 273 273 500
fix scales all print 50 "${vstep} ${distZ}" file coord-${R0}.
txt screen no
# ${distZ} is the distance between the tether point at the
surface of ice and protein center of mass
dump 1 all 1000 window-${R0}.lammpstrj id mol type xs ys zs
run 2500000
```

Ice Nucleation Simulations with the Model INP The steps to perform heterogeneous nucleation of ice is as followed:

1. Create a simulation cell by placing the united atom model of *Pseudomonas syringae* (*Ps*INP) at the center of a box of liquid mW water. Remove the water molecules within 3.5 Å of the protein to avoid bad contacts.
2. Equilibrate the system for 5 ns at 273 K in the *NVT* ensemble. Then instantly quench the system temperature to 228 K and cool the system to 198 K at a constant cooling rate of 1 K/ns. This particular cooling rate is the maximum crystallization rate for mW water [36].
3. Determine the ice nucleation temperature from the inflection point of the plot of potential energy as a function of the temperature. Compute the error bar as standard deviation from five independent cooling simulations. Figure 3 shows snapshots from a typical ice nucleation and growth simulation.

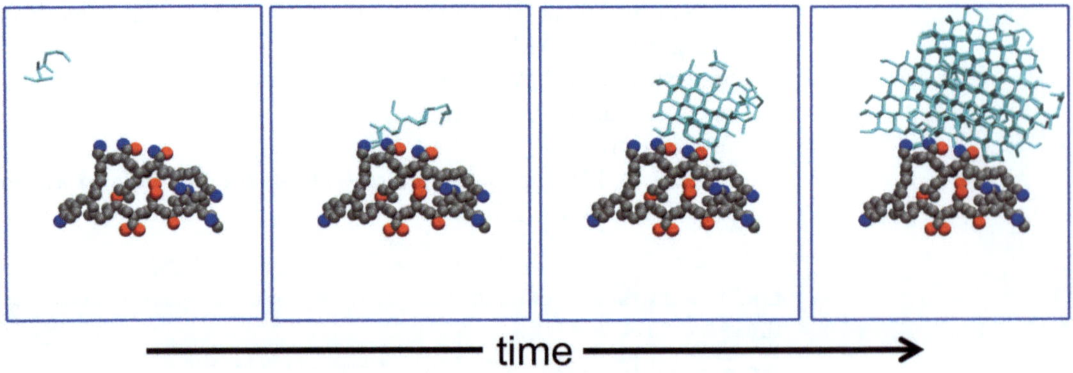

Fig. 3 Heterogeneous ice nucleation and growth of ice on model ice-nucleating protein *Ps*INP. The hydroxyl and terminal methyl groups are shown with blue and red balls, respectively. All other beads of the protein are shown in gray balls. The largest ice cluster is shown in cyan sticks. Water molecules that are not part of the largest cluster are not shown for clarity

4 Identification of Ice and Ice-Like Order

There are several order parameters that can be used to differentiate ice, ice-like order and liquid water. Steinhardt bond orientational order parameter q_l of the order $l = 3$ or 6 are widely used to recognize ice and water [60]. The bond orientational order parameter q_l is defined as

$$q_l = \left(\frac{4\pi}{2l+1} \sum_{m=-l}^{l} \left| \overline{q}_{lm} \right|^2 \right)^{1/2}$$

where $\overline{q}_{lm} = \frac{1}{N_B} \sum_{i=1}^{N_B} Y_{lm}(\theta_i(r), \phi_i(r))$ and $Y_{lm}(\theta_i(r), \phi_i(r))$ are spherical harmonics of rank l and m. $\theta_i(r)$ and $\phi_i(r)$ are the polar angles of each of the N_B bonds between the central atom i and the closest neighbors within 3.5 Å, but not exceeding maximum four neighbors. We note that when searching for the closest neighbors, both water oxygen and hydroxyl oxygen from the protein can be considered to account for the impact of the protein on the water ordering in the hydration shell. The cutoff for ice and water classification for q_3 and q_6 is as follows:

1. For the q_3 water order parameter, classify the water molecules with $q_3 > 0.6$ as ice or ice-like. The q_3 order parameter can also differentiate between anchored clathrate (AC) order of water at the ice-binding site and bulk ice. q_3 value of water molecules in the AC motif is between 0.6 and 0.68, while bulk ice is $q_3 > 0.7$.
2. Classify water molecules with $q_6 > 0.57$ as ice. To identify the largest ice cluster, use a cutoff distance of 3.5 Å between water molecules with $q_6 > 0.57$. Use q_6 order parameter with the largest ice cluster classification to identify the ice nucleus formation in the coarse-grained simulations of heterogeneous ice nucleation with the model *Ps*INP.

The user can also use tetrahedral order parameter, q_{tet}, to distinguish water and ice [62]. q_{tet} for each water molecule is computed as

$$q_{\text{tet}} = 1 - \frac{3}{8} \sum_{j=1}^{3} \sum_{k=j+1}^{4} \left(\cos\,\theta_{ijk} + \frac{1}{3} \right)^2$$

where for a central water oxygen i, four closest neighbors (j and k) are identified. The central atom (i) and the four closest neighbors form six angles θ_{ijk}. For perfect tetrahedrons where the closest four neighbors are positioned at the vertices, $\cos\,\theta_{ijk} = -1/3$, and hence $q_{\text{tet}} = 1$. In a protein–ice–water system, the central atom is always the water oxygen, and the nearest neighbors can be water oxygen or any protein atoms that can hydrogen bond with water. q_{tet} value for ice or ice-like order is typically greater than 0.7.

5 Notes

1. For the all-atom ice-binding simulations of the hyperactive insect AFPs, check the dihedral angles ($\phi_{N\text{–}C\text{–}C\text{–}OH}$ and $\phi_{N\text{–}C\text{–}C\text{–}CH3}$) of the threonine residues at the ice-binding site. Check if the hydroxyl groups are pointed inward (gauche conformation) to the channel between the two parallel rows of the threonine residues and the methyl groups are pointed outward (anti-conformation) relative to the channel.
2. In all the simulations of ice growth, it is important to construct the system such that the growing ice and hydration water of the protein do not interact with the periodic image due to the boundary conditions. A good rule of thumb is to have at least 20 Å of water from the edge of the protein to the box edge. Additionally, when estimating the curvature of the growing ice front, finite-size effects must be taken into consideration.
3. Before starting the simulations, visualize the composite ice/-water and protein system to ensure proper placement of the protein. Importantly, the ice-binding site of the protein needs to be coplanar to the plane of the ice to ensure rapid formation of bridging ice layers between the protein ice-binding site and the ice/water interface. Finally, check if there are water molecules that are overlapping with the initial configuration of the protein. Remove the bad contacts to avoid energetic instability during simulation startup.
4. Ice growth simulations are strongly influenced by temperature. At higher supercooling (further away from the melting point), the ice will grow rapidly and engulf the protein. The closer the simulation temperature is to the melting point of the water model, the slower the ice growth rate is. Hence, longer simulation times are required to achieve the integration of the ice growing front and protein-binding site. Furthermore, the slower the ice growth rate is, the higher the probability that the AFP can tumble away, resulting in misorientation of the ice-binding site relative to the ice plane. This issue can be overcome by harmonically restraining the distance between the protein geometric center and the ice surface. If this setup is used, then the protein may misorient, but it will not translate away, hence increasing the probability for the ice-binding site to achieve correct orientation relative to the ice plane.
5. The process of ice nucleation is stochastic in nature. Hence, multiple replica simulations must be performed to obtain the reliable mean and standard deviation value of the heterogeneous ice nucleation temperature. For replicate simulations to be uncorrelated, the initial water coordinates at each system must be different (even if the total number of water molecules

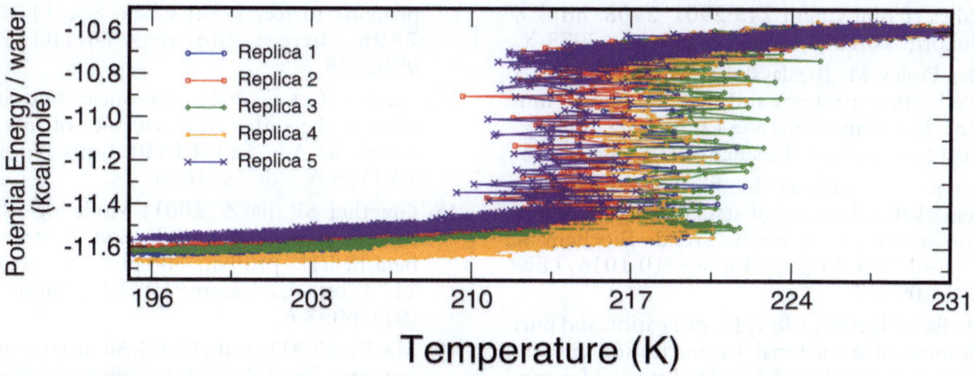

Fig. 4 Potential energy per water molecule for temperature quenching ice nucleation simulations. Note the sharp decline in the potential energy signifying the formation of a stable ice nucleus and resulting ice growth. Furthermore, the inflection point varies from 212 to 218 K indicating the stochastic nature of the ice nucleation process

and the system size is the same). Also, for each independent replica simulation, plot the potential energy against the simulation temperature to ensure there is a sharp inflection event (indicating a successful nucleation event that results in the ice growth). Figure 4 shows the potential energy profile for different replicas for a heterogeneous ice nucleation simulation mediated by model *Ps*INP.

6. For the umbrella sampling simulations, check that the histogram of the CV for two consecutive windows has significant overlap. The force constant can be adjusted to provide increased overlap between two consecutive windows. When postprocessing the CV data at each window using WHAM, exclude the first 5–10 ns of the data as equilibration time.

Acknowledgments

The author gratefully acknowledges discussions with Prof. Valeria Molinero, Dr. Yuqing Qiu, Prof. Francesco Paesani, and Dr. Daniel Moberg. The simulation trajectories were generated using the supercomputing resources and computer time provided by the Center for High Performance at the University of Utah.

References

1. McCammon JA, Gelin BR, Karplus M (1977) Dynamics of folded proteins. Nature 267:585–590. https://doi.org/10.1038/267585a0
2. Karplus M, McCammon JA (2002) Molecular dynamics simulations of biomolecules. Nat Struct Biol 9:646–652. https://doi.org/10.1038/nsb0902-646
3. DeVries Arthur L, Wohlschlag Donald E (1969) Freezing resistance in some Antarctic fishes. Science 163:1073–1075. https://doi.org/10.1126/science.163.3871.1073
4. DeVries AL, Komatsu SK, Feeney RE (1970) Chemical and physical properties of freezing point-depressing glycoproteins from Antarctic

fishes. J Biol Chem 245:2901–2908. https://doi.org/10.1016/S0021-9258(18)63073-X

5. Bar Dolev M, Braslavsky I, Davies PL (2016) Ice-binding proteins and their function. Annu Rev Biochem 85:515–542. https://doi.org/10.1146/annurev-biochem-060815-014546
6. Davies PL (2014) Ice-binding proteins: a remarkable diversity of structures for stopping and starting ice growth. Trends Biochem Sci 39:548–555. https://doi.org/10.1016/j.tibs.2014.09.005
7. Wolber PK et al (1986) Identification and purification of a bacterial ice-nucleation protein. Proc Natl Acad Sci 83:7256. https://doi.org/10.1073/pnas.83.19.7256
8. Hartmann S et al (2013) Immersion freezing of ice nucleation active protein complexes. Atmos Chem Phys 13:5751–5766. https://doi.org/10.5194/acp-13-5751-2013
9. Liou Y-C, Tocilj A, Davies PL, Jia Z (2000) Mimicry of ice structure by surface hydroxyls and water of a β-helix antifreeze protein. Nature 406:322–324. https://doi.org/10.1038/35018604
10. Leinala EK et al (2002) A β-helical antifreeze protein isoform with increased activity: structural and functional insights*. J Biol Chem 277:33349–33352. https://doi.org/10.1074/jbc.M205575200
11. Hakim A et al (2013) Crystal structure of an insect antifreeze protein and its implications for ice binding*. J Biol Chem 288:12295–12304. https://doi.org/10.1074/jbc.M113.450973
12. Garnham CP, Campbell RL, Davies PL (2011) Anchored clathrate waters bind antifreeze proteins to ice. Proc Natl Acad Sci 108:7363. https://doi.org/10.1073/pnas.1100429108
13. Hudait A, Odendahl N, Qiu Y, Paesani F, Molinero V (2018) Ice-nucleating and antifreeze proteins recognize ice through a diversity of anchored clathrate and ice-like motifs. J Am Chem Soc 140:4905–4912. https://doi.org/10.1021/jacs.8b01246
14. Hudait A et al (2018) Preordering of water is not needed for ice recognition by hyperactive antifreeze proteins. Proc Natl Acad Sci 115:8266. https://doi.org/10.1073/pnas.1806996115
15. Qiu Y, Hudait A, Molinero V (2019) How size and aggregation of ice-binding proteins control their ice nucleation efficiency. J Am Chem Soc 141:7439–7452. https://doi.org/10.1021/jacs.9b01854
16. Hudait A, Qiu Y, Odendahl N, Molinero V (2019) Hydrogen-bonding and hydrophobic groups contribute equally to the binding of hyperactive antifreeze and ice-nucleating proteins to ice. J Am Chem Soc 141:7887–7898. https://doi.org/10.1021/jacs.9b02248
17. Pandey R et al (2016) Ice-nucleating bacteria control the order and dynamics of interfacial water. Sci Adv 2:e1501630. https://doi.org/10.1126/sciadv.1501630
18. Graether SP, Jia Z (2001) Modeling pseudomonas syringae ice-nucleation protein as a beta-helical protein. Biophys J 80:1169–1173. https://doi.org/10.1016/S0006-3495(01)76093-6
19. MacKerell AD et al (1998) All-atom empirical potential for molecular modeling and dynamics studies of proteins. J Phys Chem B 102:3586–3616. https://doi.org/10.1021/jp973084f
20. MacKerell AD, Feig M, Brooks CL (2004) Improved treatment of the protein backbone in empirical force fields. J Am Chem Soc 126: 698–699. https://doi.org/10.1021/ja036959e
21. Meister K et al (2015) Investigation of the ice-binding site of an insect antifreeze protein using sum-frequency generation spectroscopy. J Phys Chem Lett 6:1162–1167. https://doi.org/10.1021/acs.jpclett.5b00281
22. Odendahl N (2016) Comparison of popular force fields for molecular modelling of proteins applied to ice binding Tenebrio Molitor antifreeze proteins. B.S. in Chemistry Honors thesis, The University of Utah, Salt Lake City
23. Abascal JLF, Vega C (2005) A general purpose model for the condensed phases of water: TIP4P/2005. J Chem Phys 123:234505. https://doi.org/10.1063/1.2121687
24. Babin V, Leforestier C, Paesani F (2013) Development of a "first principles" water potential with flexible monomers: dimer potential energy surface, VRT spectrum, and second virial coefficient. J Chem Theory Comput 9:5395–5403. https://doi.org/10.1021/ct400863t
25. Babin V, Medders GR, Paesani F (2014) Development of a "first principles" water potential with flexible monomers. II: trimer potential energy surface, third virial coefficient, and small clusters. J Chem Theory Comput 10: 1599–1607. https://doi.org/10.1021/ct500079y
26. Medders GR, Babin V, Paesani F (2014) Development of a "first-principles" water potential with flexible monomers. III. Liquid phase properties. J Chem Theory Comput 10: 2906–2910. https://doi.org/10.1021/ct5004115
27. Horn HW et al (2004) Development of an improved four-site water model for

biomolecular simulations: TIP4P-Ew. J Chem Phys 120:9665–9678. https://doi.org/10.1063/1.1683075

28. Berendsen HJC, Grigera JR, Straatsma TP (1987) The missing term in effective pair potentials. J Phys Chem 91:6269–6271. https://doi.org/10.1021/j100308a038
29. Abascal JLF, Sanz E, García Fernández R, Vega C (2005) A potential model for the study of ices and amorphous water: TIP4P/ice. J Chem Phys 122:234511. https://doi.org/10.1063/1.1931662
30. Case DA et al (2005) The Amber biomolecular simulation programs. J Comput Chem 26:1668–1688. https://doi.org/10.1002/jcc.20290
31. Berendsen HJC, van der Spoel D, van Drunen R (1995) GROMACS: a message-passing parallel molecular dynamics implementation. Comput Phys Commun 91:43–56. https://doi.org/10.1016/0010-4655(95)00042-E
32. Phillips JC et al (2020) Scalable molecular dynamics on CPU and GPU architectures with NAMD. J Chem Phys 153:044130. https://doi.org/10.1063/5.0014475
33. Smith W, Yong CW, Rodger PM (2002) DL_POLY: application to molecular simulation. Mol Simul 28:385–471. https://doi.org/10.1080/08927020290018769
34. Thompson AP et al (2022) LAMMPS - a flexible simulation tool for particle-based materials modeling at the atomic, meso, and continuum scales. Comput Phys Commun 271:108171. https://doi.org/10.1016/j.cpc.2021.108171
35. http://paesanigroup.ucsd.edu/software/mbx.html
36. Moore EB, Molinero V (2011) Structural transformation in supercooled water controls the crystallization rate of ice. Nature 479:506–508. https://doi.org/10.1038/nature10586
37. Lupi L et al (2017) Role of stacking disorder in ice nucleation. Nature 551:218–222. https://doi.org/10.1038/nature24279
38. Hudait A, Molinero V (2014) Ice crystallization in ultrafine water–salt aerosols: nucleation, ice-solution equilibrium, and internal structure. J Am Chem Soc 136:8081–8093. https://doi.org/10.1021/ja503311r
39. Metya AK, Molinero V (2021) Is ice nucleation by organic crystals nonclassical? An assessment of the monolayer hypothesis of ice nucleation. J Am Chem Soc 143:4607–4624. https://doi.org/10.1021/jacs.0c12012
40. Hudait A, Molinero V (2016) What determines the ice polymorph in clouds? J Am Chem Soc 138:8958–8967. https://doi.org/10.1021/jacs.6b05227
41. Qiu Y, Molinero V (2018) Why is it so difficult to identify the onset of ice Premelting? J Phys Chem Lett 9:5179–5182. https://doi.org/10.1021/acs.jpclett.8b02244
42. Hudait A, Qiu S, Lupi L, Molinero V (2016) Free energy contributions and structural characterization of stacking disordered ices. Phys Chem Chem Phys 18:9544–9553. https://doi.org/10.1039/C6CP00915H
43. Molinero V, Moore EB (2009) Water modeled as an intermediate element between carbon and silicon. J Phys Chem B 113:4008–4016. https://doi.org/10.1021/jp805227c
44. Humphrey W, Dalke A, Schulten K (1996) VMD: visual molecular dynamics. J Mol Graph 14:33–38. https://doi.org/10.1016/0263-7855(96)00018-5
45. Berendsen HJC, Postma JPM, van Gunsteren WF, DiNola A, Haak JR (1984) Molecular dynamics with coupling to an external bath. J Chem Phys 81:3684–3690. https://doi.org/10.1063/1.448118
46. Schneider T, Stoll E (1978) Molecular-dynamics study of a three-dimensional one--component model for distortive phase transitions. Phys Rev B 17:1302–1322. https://doi.org/10.1103/PhysRevB.17.1302
47. Grønbech-Jensen N, Farago O (2013) A simple and effective Verlet-type algorithm for simulating Langevin dynamics. Mol Phys 111:983–991. https://doi.org/10.1080/00268976.2012.760055
48. Ryckaert J-P, Ciccotti G, Berendsen HJC (1977) Numerical integration of the cartesian equations of motion of a system with constraints: molecular dynamics of n-alkanes. J Comput Phys 23:327–341. https://doi.org/10.1016/0021-9991(77)90098-5
49. Andersen HC (1983) Rattle: a "velocity" version of the shake algorithm for molecular dynamics calculations. J Comput Phys 52:24–34. https://doi.org/10.1016/0021-9991(83)90014-1
50. Darden T, York D, Pedersen L (1993) Particle mesh Ewald: an N·log(N) method for Ewald sums in large systems. J Chem Phys 98:10089–10092. https://doi.org/10.1063/1.464397
51. Kolafa J (2004) Time-reversible always stable predictor–corrector method for molecular dynamics of polarizable molecules. J Comput Chem 25:335–342. https://doi.org/10.1002/jcc.10385
52. Parrinello M, Rahman A (1981) Polymorphic transitions in single crystals: a new molecular

dynamics method. J Appl Phys 52:7182–7190. https://doi.org/10.1063/1.328693

53. Martyna GJ, Tobias DJ, Klein ML (1994) Constant pressure molecular dynamics algorithms. J Chem Phys 101:4177–4189. https://doi.org/10.1063/1.467468
54. Stillinger FH, Weber TA (1985) Computer simulation of local order in condensed phases of silicon. Phys Rev B 31:5262–5271. https://doi.org/10.1103/PhysRevB.31.5262
55. Fileti EE, Chaudhuri P, Canuto S (2004) Relative strength of hydrogen bond interaction in alcohol–water complexes. Chem Phys Lett 400:494–499. https://doi.org/10.1016/j.cplett.2004.10.149
56. Jacobson LC, Hujo W, Molinero V (2010) Amorphous precursors in the nucleation of clathrate hydrates. J Am Chem Soc 132: 11806–11811. https://doi.org/10.1021/ja1051445
57. Kamberaj H, Low RJ, Neal MP (2005) Time reversible and symplectic integrators for molecular dynamics simulations of rigid molecules. J Chem Phys 122:224114. https://doi.org/10.1063/1.1906216
58. Miller TF et al (2002) Symplectic quaternion scheme for biophysical molecular dynamics. J Chem Phys 116:8649–8659. https://doi.org/10.1063/1.1473654
59. Torrie GM, Valleau JP (1977) Nonphysical sampling distributions in Monte Carlo free-energy estimation: umbrella sampling. J Comput Phys 23:187–199. https://doi.org/10.1016/0021-9991(77)90121-8
60. Steinhardt PJ, Nelson DR, Ronchetti M (1983) Bond-orientational order in liquids and glasses. Phys Rev B 28:784–805. https://doi.org/10.1103/PhysRevB.28.784
61. Grossfield A, WHAM; version 2.0.6, http://membrane.urmc.rochester.edu/content/wham
62. Errington JR, Debenedetti PG (2001) Relationship between structural order and the anomalies of liquid water. Nature 409:318–321. https://doi.org/10.1038/35053024

Chapter 14

Synthesis of Polymeric Mimics of Ice-Binding Proteins

Christian C. M. Sproncken, Christophe Detrembleur, and Ilja K. Voets

Abstract

Cobalt-mediated radical polymerization (CMRP) enables the preparation of both short and long polymers from acrylic and vinyl ester monomers with low dispersity. Here we describe the synthesis, purification, and characterization of polymeric mimics of ice-binding proteins based on the water-soluble polymer poly(vinyl alcohol) by CMRP. Block copolymers of poly(vinyl alcohol) and poly(acrylic acid) were prepared from the precursor copolymers poly(vinyl acetate)-*b*-poly(acrylonitrile) upon hydrolysis. Copolymers comprising up to hundreds of monomers and dispersities $M_w/M_n < 1.3$ were produced by this method.

Key words Cobalt-mediated radical polymerization, Controlled radical polymerization, Poly(vinyl alcohol), Poly(vinyl alcohol)-*b*-poly(acrylic acid), Polymeric mimic of ice-binding proteins, Polymeric antifreeze, Size exclusion chromatography, ^{1}H NMR spectroscopy

1 Introduction

Controlled radical polymerization (CRP) techniques offer access to a broad range of polymers customized in composition, dimensions, and chain architecture. Since the discovery of CRP in 1993 [1], various methods have been established, including nitroxide-mediated radical polymerization (NMP) [2, 3], atom transfer radical polymerization (ATRP) [4], radical addition–fragmentation chain transfer (RAFT) [5], and cobalt-mediated radical polymerization (CMRP) [6, 7]. Most polymeric ice-binding protein (IBP) mimics are based on the water-soluble polymer poly(vinyl alcohol) (PVOH), which is typically prepared by CMRP [8] or RAFT polymerization [9, 10] of vinyl acetate followed by hydrolysis of the prepared poly(vinyl acetate). Above a sufficiently high molecular weight, PVOH is able to inhibit ice recrystallization, i.e., the polymer displays ice recrystallization inhibition (IRI) activity. PVOH is the most IRI active polymer known to date with an inhibitory concentration c_i of 0.3–0.4 mM vinyl alcohol (corresponding to 0.013–0.018 g L^{-1}) above which ice crystal recrystallization rates drop markedly. Compared to natural IBPs, PVOH is moderately

Ran Drori and Corey A. Stevens (eds.), *Ice Binding Proteins: Methods and Protocols*, Methods in Molecular Biology, vol. 2730, https://doi.org/10.1007/978-1-0716-3503-2_14,

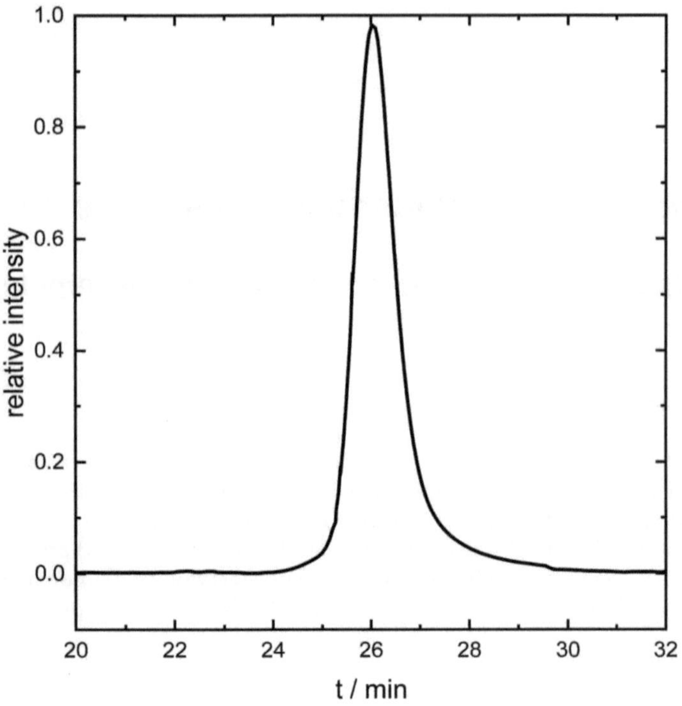

Fig. 1 SEC chromatogram of the PVAc macroinitiator in DMF with 0.025 M LiBr

active. Interestingly, the IRI activity of PVOH is little affected by adjustments in the composition of the polymer, as long as sufficiently long stretches of vinyl alcohol monomers remain included. This offers ample room to customize polymer composition, architecture, and solution behavior to develop IRI active materials with interesting and diverse (auxiliary) properties. Here we describe the synthesis, purification, and characterization of the block copolymer poly(vinyl alcohol)-*b*-poly(acrylic acid) (PVOH-*b*-PAA), which has been used to prepare IRI active polymer micelles [8]. The procedure for the preparation of PVOH-*b*-PAA has been reported previously in [11]. PVOH-containing polymers varying in composition and architecture have been achieved in a similar fashion [12–14]. Polymers with hundreds of monomers and dispersities $M_w/M_n < 1.3$ were prepared as demonstrated by size exclusion chromatography (SEC) (*see* Figs. 1 and 2) and ^{1}H NMR spectroscopy (*see* Figs. 3 and 4).

2 Materials

1. Vinyl acetate (>99%, Aldrich). Dry over calcium hydride. Degas by several freeze–thawing cycles. Distill under reduced pressure. Store under argon.

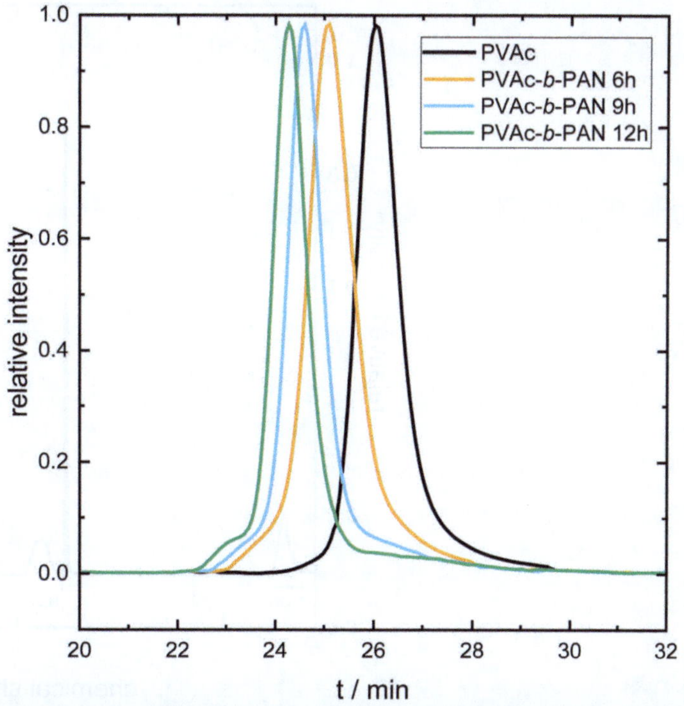

Fig. 2 SEC chromatogram of the PVAc-*b*-PAN precursors, after 6 (yellow), 9 (blue), and 12 h (green) of polymerization, compared to the trace of the PVAc macroinitiator (black), in DMF with 0.025 M LiBr

2. Acrylonitrile (>99%, Aldrich). Dry over calcium hydride. Degas by several freeze–thawing cycles. Distill under reduced pressure. Store under argon.
3. 2,2′-Azobis(4-methoxy-2,4-dimethylvaleronitrile) (V-70) (Wako).
4. Cobalt(II) acetylacetonate ($Co(acac)_2$) (>98%, Merck).
5. 2,2,6,6-Tetramethylpiperidine-1-oxyl (TEMPO) (98%, Aldrich).
6. 500 mL flask.
7. 5 L flask.
8. 3 L flask.
9. Plastic syringe.
10. Nitrogen gas.
11. Tetrahydrofuran (THF).
12. Dimethylformamide (DMF) with 0.025 M LiBr.
13. Methanol.
14. Ethanol (p.a.).
15. DMSO-d_6.
16. KOH.
17. Deionized water.

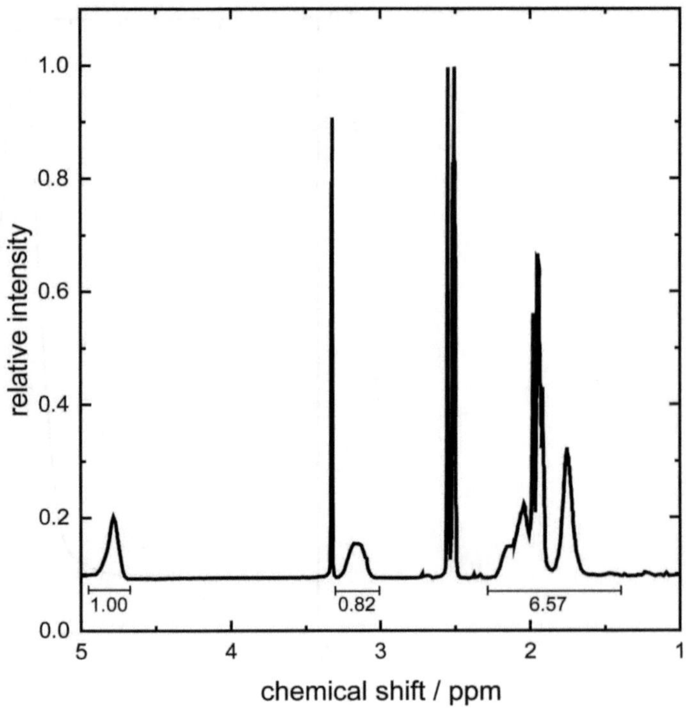

Fig. 3 ^{1}H NMR spectrum of the precursor PVAc-*b*-PAN in DMSO-d_6. δ (ppm): 3.14 (2H, CH_2-<u>CH</u>-CN of PAN), 4.78 (2H, CH_2-<u>CH</u>-$OCOCH_3$ of PVAc)

18. Spectra/Pore membrane (cutoff: 6000–8000 Da).
19. Polystyrene standards ranging from 225 to 580 × 10^4 g/mol for SEC calibration.
20. Waters 600 liquid chromatograph equipped with a 410 refractive index detector and Styragel HR columns.
21. 250 MHz Bruker spectrometer.

3 Methods

3.1 Synthesis, Purification, and Characterization of the Poly(vinyl acetate)-b-poly(acrylonitrile) Precursor PVAc-b-PAN

1. Add Co(acac)$_2$ (1.03 g, 4.02 mmol) and V-70 (4.01 g, 13.0 mmol) in a 500 mL flask.
2. Degas the mixture three times by vacuum–nitrogen cycles.
3. Add vinyl acetate (150 mL, 1.62 mol) with a syringe under nitrogen.
4. Stir the purple mixture at 30 °C for 26 h.
5. After 26 h, withdraw a sample for characterization: evaluate the monomer conversion gravimetrically, and measure the molar mass and dispersity of the PVAc macroinitiator by SEC analysis in DMF with 0.025 M LiBr (*see* Fig. 1).

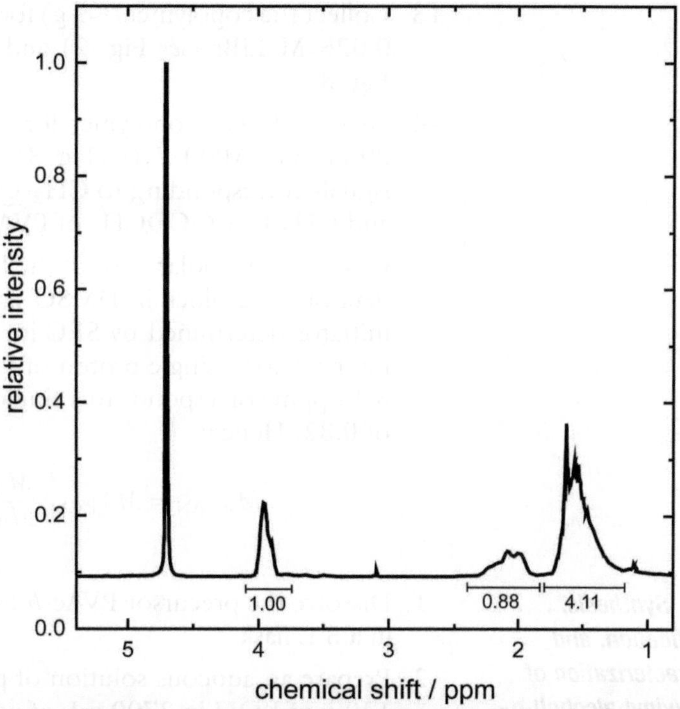

Fig. 4 ^{1}H NMR spectrum of PVOH-*b*-PAA in D_2O. δ (ppm): 3.9 (1H, CH-OH of PVOH), 3.09 (3H, end group), 2.05 (1H, CH-COO-K+ of PAA), 1.80–1.0 (2H, CH_2-CHOH of PVOH and 2H, CH_2-CH-COO-K+ of PAA)

6. Evaporate the residual monomer under reduced pressure at room temperature.
7. Store the pink poly(vinyl acetate) macroinitiator under inert atmosphere.
8. Cool the flask to 0 °C in an ice bath and add 30 mL of dry DMSO.
9. Add 25 mL of acrylonitrile (AN).
10. Stir the reaction medium at 0 °C for about 12 h to polymerize AN. Remove 10 mL of the reaction mixture after 6 h and another 10 mL after 9 h. During AN polymerization, the medium became viscous but remained homogeneous, and the color changed from pink to gray black.
11. Add the two 10 mL aliquots and the final reaction mixture to a solution of TEMPO (50 mg in 5 mL of DMSO).
12. Disperse the copolymer in a methanol/water (20/80) mixture from which it will precipitate. Filter the copolymer using a paper filter and wash it with more MeOH/H_2O. Dry the copolymer under vacuum at 80 °C.

13. Collect the copolymer (~9 g) for analysis by SEC in DMF with 0.025 M LiBr (*see* Fig. 2) and ^1H NMR in DMSO-d_6 (*see* Fig. 3).
14. Dissolve 5 wt% copolymer for ^1H NMR analysis (250 MHz, 298 K) in DMSO-d_6 (*see* Fig. 3). Compare the intensities of the signals corresponding to CH_2-<u>CH</u>-CN of PAN (at 3.14 ppm) and CH_2-<u>CH</u>-$OCOCH_3$ of PVAc (at 4.78 ppm).
15. Calculate the molar mass of the PAN block from the ^1H NMR data of the diblock in DMSO-d_6 and M_n of the PVAc macroinitiator determined by SEC in THF. Having normalized the intensity for a single proton of PVAc at 4.78 ppm, the peak at 3.14 ppm corresponds to a single proton of PAN, with an area of 0.82. Hence:

$$M_{n,PAN} = M_{n,PVAc} \cdot \frac{M_{0,PAN}}{M_{0,PVAc}} \cdot \frac{A_{3.14ppm}}{A_{4.78ppm}}$$

3.2 Synthesis, Purification, and Characterization of Poly(vinyl alcohol)-b-poly(acryl acid) (PVOH-b-PAA)

1. Dissolve the precursor PVAc-*b*-PAN (30 g) in 600 mL ethanol in a 5 L flask.
2. Prepare an aqueous solution of potassium hydroxide: dissolve 120 g of KOH in 2700 mL of water in a 3 L flask.
3. Add the aqueous solution of potassium hydroxide to the precursor solution in ethanol.
4. Stir the mixture at 75 °C for 48 h under reflux. Over the course of hours, the copolymer was sufficiently hydrolyzed to slowly dissolve in the reaction medium to form a homogeneous solution.
5. Remove ethanol under vacuum.
6. Dialyze the aqueous copolymer solution for 48 h against water through a Spectra/Pore membrane (cutoff: 6000–8000 Da).
7. Lyophilize the copolymer.
8. Recover the copolymer as a white powder.
9. Dissolve 5 wt% copolymer in D_2O for ^1H NMR analysis (250 MHz, 298 K) (*see* Fig. 4).
10. Calculate the molar mass of the PVOH block from the ^1H NMR data of the diblock in D_2O.

$$M_{n,POH} = 3 \cdot M_{0,PVAc} \cdot \frac{A_{3.90ppm}}{A_{3.09ppm}}$$

The molar ratio of the PVOH/PAA blocks calculated from the ^1H NMR spectrum of PVOH-*b*-PAA was very similar to the molar ratio of the PVAc/PAN blocks of the PVAc-*b*-PAN precursor.

$$M_{n,PAA} = M_{n,PVOH} \cdot \frac{M_{0,PAA}}{M_{0,PVOH}} \cdot \frac{A_{2.05ppm}}{A_{3.90ppm}}$$

3.3 Size Exclusion Chromatography (SEC)

1. Perform size exclusion chromatography (SEC) analysis on a Waters 600 liquid chromatograph equipped with a 410 refractive index detector and Styragel HR columns. Use polystyrene (PS) standards with molar masses ranging from 225 to 580×10^4 g/mol for calibration.
2. For SEC analysis of the molar mass and dispersity of the PVAc macroinitiator and PVAc-*b*-PAN precursor in DMF with 0.025 M LiBr (*see* Fig. 1): use a flow rate of 1 mL min^{-1} and a temperature of 55 °C. Use Styragel HR columns (HR1, 100–5000; HR3, 500–30,000; HR4, 5000–500,000; HR5, 2000–40,000,000). Use PS standards for calibration (*see* **Note 1**).

3.4 ^{1}H NMR Spectroscopy

1. Record ^{1}H NMR spectra on a 250 MHz Bruker spectrometer running at 298 K.
2. For ^{1}H NMR analysis of the PVAc macroinitiator: Dissolve 5 wt% PVAc in $CDCl_3$ and record 16 scans with $D_1 = 2$ s.
3. For ^{1}H NMR analysis of the precursor PVAc-*b*-PAN: Dissolve 5 wt% of the precursor in DMSO-d_6 (*see* Fig. 3) and record 32 scans with $D_1 = 5$ s.
4. For ^{1}H NMR analysis of the copolymer PVOH-*b*-PAA: Dissolve 5 wt% of the copolymer in D_2O (*see* Fig. 4) and record 32 scans with $D_1 = 5$ s.

4 Notes

1. The molar mass determination of the PVAc macroinitiator by SEC analysis using PS standards and ^{1}H NMR in $CDCl_3$ was in good agreement whenever the -OCH_3 protons of the initiator at $\delta = 3.13$ ppm could be observed and compared to the -$\underline{CH}OCOCH_3$ proton of the vinyl acetate monomer unit at $\delta = 4.8$ ppm as reported elsewhere previously [15, 16].

Acknowledgments

This work is financially supported by the European Union (ERC-2020-CoG contract no. 101001965) and the Dutch Science Foundation (NWO ECHO grant no. 712.016.002). C.D. is Research Director of F.R.S.-FNRS and thanks FNRS (Belgium) for financial support.

References

1. Georges MK, Veregin RPN, Kazmaier PM, Hamer GK (1993) Narrow molecular weight resins by a free-radical polymerization process. Macromolecules 26:2987–2988. https://doi.org/10.1021/ma00063a054
2. Hawker CJ, Bosman AW, Harth E (2001) New polymer synthesis by nitroxide mediated living radical polymerizations. Chem Rev 101:3661–3688. https://doi.org/10.1021/cr990119u
3. Sciannamea V, Jérôme R, Detrembleur C (2008) In-situ nitroxide-mediated radical polymerization (NMP) processes: their understanding and optimization. Chem Rev 108:1104–1126. https://doi.org/10.1021/cr0680540
4. Matyjaszewski K, Xia J (2010) Atom transfer radical polymerization. Prog Chem 22:2079–2088. https://doi.org/10.1002/9783527809080.cataz01278
5. Perrier S, Takolpuckdee P (2005) Macromolecular design via reversible addition-fragmentation chain transfer (RAFT)/xanthates (MADIX) polymerization. J Polym Sci Part A Polym Chem 43:5347–5393. https://doi.org/10.1002/pola.20986
6. Debuigne A, Poli R, Jérôme C et al (2009) Overview of cobalt-mediated radical polymerization: roots, state of the art and future prospects. Prog Polym Sci 34:211–239. https://doi.org/10.1016/j.progpolymsci.2008.11.003
7. Demarteau J, Debuigne A, Detrembleur C (2019) Organocobalt complexes as sources of carbon-centered radicals for organic and polymer chemistries. Chem Rev 119:6906–6955. https://doi.org/10.1021/acs.chemrev.8b00715
8. Sproncken CCM, Surís-Valls R, Cingil HE et al (2018) Complex coacervate core micelles containing poly(vinyl alcohol) inhibit ice recrystallization. Macromol Rapid Commun. https://doi.org/10.1002/marc.201700814
9. Congdon TR, Notman R, Gibson MI (2017) Synthesis of star-branched poly(vinyl alcohol) and ice recrystallization inhibition activity. Eur Polym J 88:320–327. https://doi.org/10.1016/j.eurpolymj.2017.01.039
10. Vail NS, Stubbs C, Biggs CI, Gibson MI (2017) Ultralow dispersity poly(vinyl alcohol) reveals significant dispersity effects on ice recrystallization inhibition activity. ACS Macro Lett 1001–1004. https://doi.org/10.1021/acsmacrolett.7b00595
11. Debuigne A, Warnant J, Jérôme R et al (2008) Synthesis of novel well-defined poly(vinyl acetate)-b-poly(acrylonitrile) and derivatized water-soluble poly(vinyl alcohol)-b-poly(acrylic acid) block copolymers by cobalt-mediated radical polymerization. Macromolecules 41. https://doi.org/10.1021/ma702341v
12. Wang Z, Vertruyen B, Taghipour H et al (2021) CO2-derived methylene oxazolidinone: a platform building block for functionalizing ethylene-vinyl alcohol copolymers. Macromolecules 54:10415–10427. https://doi.org/10.1021/acs.macromol.1c01895
13. Wang Z, Poli R, Detrembleur C, Debuigne A (2019) Organometallic-mediated radical (Co)-polymerization of γ-methylene-γ-butyrolactone: access to pH-responsive poly(vinyl alcohol) derivatives. Macromolecules 52:8976–8988. https://doi.org/10.1021/acs.macromol.9b01838
14. Scholten PBV, Demarteau J, Gennen S et al (2018) Merging CO2-based building blocks with cobalt-mediated radical polymerization for the synthesis of functional poly(vinyl alcohol)s. Macromolecules 51:3379–3393. https://doi.org/10.1021/acs.macromol.8b00492
15. Debuigne A, Caille J-R, Detrembleur C, Jérôme R (2005) Effective cobalt mediation of the radical polymerization of vinyl acetate in suspension. Angew Chemie Int Ed 44:3439–3442. https://doi.org/10.1002/anie.200500112
16. Debuigne A, Caille JR, Jerome R (2005) Highly efficient cobalt-mediated radical polymerization of vinyl acetate. Angew Chemie Int Ed 44:1101–1104. https://doi.org/10.1002/anie.200461333

Chapter 15

Generating Ice-Binding Protein–Polymer Bioconjugates

Corey A. Stevens

Abstract

Protein–polymer conjugates combine the stability of polymers with the diversity, specificity, and functionality of proteins. The resulting hybrid materials can display properties not found in the individual components and can be particularly relevant for engineering new functionalities. Ice-binding proteins have many potential biotechnical and biomedical applications. However, their widespread use has been limited due to cost of production, limited activity, and relative instability. Polymer attachment has led to higher thermal hysteresis activities with less protein and superior stabilities. Thus, IBP–polymer conjugates have the ability to overcome a number of these challenges and lead to materials to tackle biomedical applications.

Key words Protein, polymer conjugates, Hybrid materials, Polymer attachment, Ice-binding proteins (IBPs), Biotechnical applications, Biomedical applications, Thermal hysteresis

1 Introduction

1.1 Protein–Polymer Conjugates

Protein–polymer bioconjugate materials offer interesting opportunities to combine the wide functional range of biomolecules with the convenient handling, processability, and stability of polymers [1, 2]. The attachment of multiple proteins to a polymer could enhance their collective behavior, through multivalent effects, and polymer attachment could also lead to improvements in solubility and stability. For instance, PEGylation, the attachment of polyethylene glycol (PEG) to a protein, has been shown to increase protein stability and blood stream circulation leading to improved pharmaceutical efficacy [3]. Through steric repulsion, PEGylation can prevent proteolysis and shield antigenic groups leading to lower immunogenicity and improved stability. PEG–protein bioconjugates have had a large clinical impact, leading to treatments for a variety of diseases including chronic hepatitis C and rheumatoid arthritis. Researchers have adopted and applied this strategy to ice-binding proteins to generate new ice-active materials.

Ran Drori and Corey A. Stevens (eds.), *Ice Binding Proteins: Methods and Protocols*, Methods in Molecular Biology, vol. 2730, https://doi.org/10.1007/978-1-0716-3503-2_15,

1.2 Ice-Binding Protein–Polymer Conjugates

Ice-binding proteins (IBPs) are found in every kingdom of life and rely on key amino acid residues to bind the surfaces of small ice crystals, preventing further growth at temperatures within a characteristic thermal hysteresis range [4]. These proteins also inhibit the subsequent growth of small crystals into larger ice crystals. Organisms inhabiting subfreezing environments have evolved IBPs to slow the growth of ice crystal embryos that could otherwise cause tissue damage. The unique ability of IBPs to depress the freezing point of water and prevent freezing has been exploited for use in food additives, and their protective properties are currently being investigated for tissue cryopreservation, vaccine storage, low-temperature surgeries, and the inhibition of gas clathrate formation [4]. IBPs have been attached to bottlebrush and dendritic polymers leading to improved activities and stabilities and new ice-active materials [5–7].

1.3 Available Chemistries for Generating Protein–Polymer Bioconjugates

There are a variety of chemistries to accommodate the attachment of a protein to a polymer resulting in a bioconjugate. Ideally, the attachment chemistry should be specific, efficient, and mild to maintain both protein structure and function and avoid the key functional residues. Successful conjugation of a protein to a polymer requires a chemically accessible reactive group present on the protein, termed reactive protein handles. Here, we will briefly describe commonly used reaction chemistries. Figure 1 provides a summary of commonly targeted reactive protein handles and reactive groups.

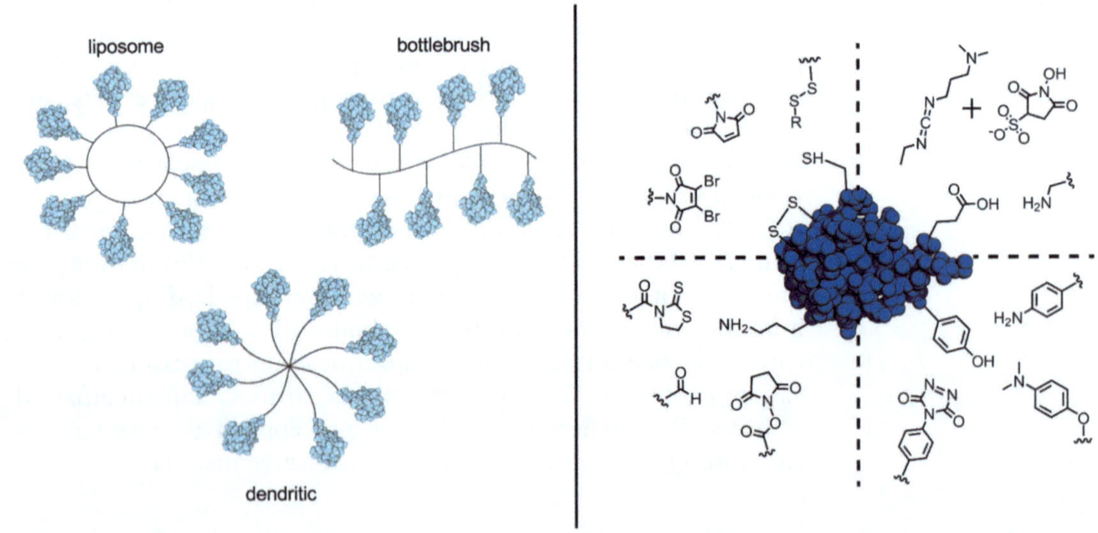

Fig. 1 (Left) Select protein–polymer bioconjugate architectures. (Right) Common protein handles/reactive groups for covalent polymer bioconjugates. Various chemistries are available to conjugate polymers to thiols, amines, tyrosine, and carboxylic acid chemistries. Adapted with permission [5]

A frequent target for protein modification is surface accessible primary amines, present at the N terminus or lysine side chains [8–10]. In fact, many of the approved PEGylated protein therapeutics target amines. Lysine is a basic amino acid relatively abundant on the protein surface making them solvent exposed and accessible for modification. However, due to the ubiquity and prevalence of lysine, it is rarely chosen as a target for site-specific modification. For site-specific bioconjugation, cysteine is a popular target due to its relatively low abundance and high nucleophilicity, allowing for relatively straightforward modification by various reactive groups including maleimide and iodoacetamide [8–10]. However, cysteines are often paired in disulfide bonds that are integral to protein folding, structure, or function and thus unavailable for modification. Due to advances in molecular biology and protein engineering, free cysteine residues can be incorporated into precise locations in a protein for specific polymer conjugation and to control protein orientation. Other targets include tyrosine and carboxylic acids found on glutamic acid, aspartic acid, and the C terminus. However, the high relative abundance of these handles on the protein surface limits conjugation specificity, and as a result, these targets have yet to be widely adopted for bioconjugate synthesis. To facilitate site-specific conjugation, both canonical and non-canonical amino acids can be incorporated at desired locations via bioengineering. Non-canonical amino acids with functional groups orthogonal to canonical amino acids represent a vast toolbox for site-specific and residue-specific formation of polymer bioconjugates [8–10]. The library of non-canonical amino acids available to conjugate polymers to proteins contains a variety of different reactive groups including those that facilitate "click" chemistry and initiators for living polymerization. Finally, a wide variety of fusion tags have been developed and used to specifically conjugate polymers to proteins. The fusion tags can vary in reactivity and chain length and be designed to have self-cleaving properties like inteins [11]. In sum, there exist a variety of strategies to from protein–polymer bioconjugates, and the repertoire continues to expand. Each synthesis strategy must weigh the benefits and drawbacks before deciding the approach is best suited for the end application. Here we will outline the method for one common protein–polymer conjugation reaction.

2 Materials

2.1 Consumable Materials

1. Purified cysteine-containing ice-binding protein at millimolar concentrations in phosphate-buffered saline (PBS pH = 7.4).
2. Polyamidoamine (PAMAM) dendrimer (Dendritech).
3. $SM(PEG)_n$ cross-linker (Thermo Fisher).

4. Sephadex PD-10 desalting spin columns (Sigma).
5. Centricon 3 kDa molecular weight cutoff centrifugal filter (GE Healthcare).
6. 15 mL conical tubes.
7. 1.5 mL microfuge tubes.
8. Gel permeation chromatography G-75 column (Amersham Biosciences).
9. Dithiothreitol (DTT) or alternative reducing agent (TCEP or β-propionic acid).
10. 10% SDS-PAGE gel.
11. SDS running and loading buffers.

2.2 Research Equipment

1. AKTA FPLC (or alternative purification systems).
2. MALDI-TOF.
3. SDS-PAGE apparatus.
4. Nanoliter osmometer (*see* Chap. 5).
5. Ice recrystallization inhibition assay (*see* Chap. 6).

3 Methods

3.1 Conjugating IBP to Polymer

1. Reduce ~30 mg of cysteine-containing protein in 1X PBS pH = 7.4 and 5 mM DTT for >2 h.
2. Dilute dendrimer (or polymer) to 30 μM in buffer (50 mM Hepes (pH = 8.4), 50 mM NaCl).
3. Add 20 μL of 250 mM SM(PEG) to dendrimer-containing solution.
4. Incubate at room temperature for 1 h.
5. Remove unreacted cross-linker via PD-10 desalting column (Sigma) following manufacturer's instructions.
6. Remove DTT from ice-binding protein solution via PD-10 desalting column or 3 kDa Centricon.
7. Incubate at room temperature for 4 h.
8. Assess conjugation reaction by SDS-PAGE (*see* Fig. 2 and **Note 1**).

3.2 Physical Characterization of Protein–Polymer Conjugate

3.2.1 Size-Exclusion Chromatography Purification of Conjugate

Perform size-exclusion purification using FPLC (Sephadex S200 HiPrep™) on the protein–polymer conjugate and protocol outlined by the manufacturer (*see* **Note 2**).

1. Attach the size-exclusion column (HiPrep™ Sephadex S200) drop to drop to the FPLC system equipped with a sample injection loop of 1 mL, and monitor at 215 and 280 nm wavelengths.

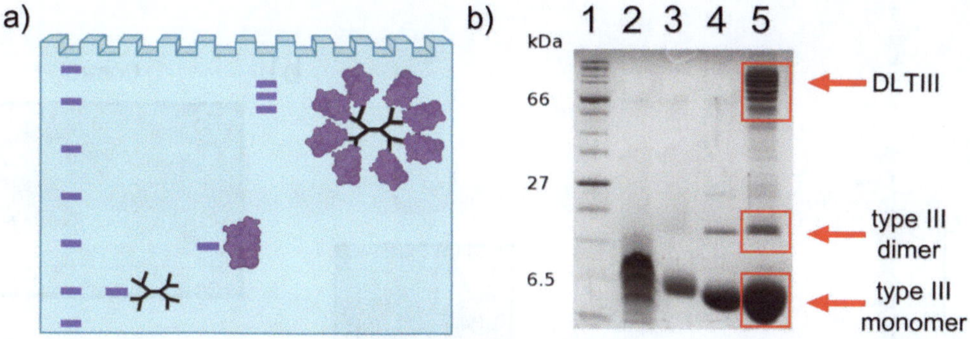

Fig. 2 Assessing protein–polymer bioconjugate formation by SDS PAGE. (**a**) Schematic representation of SDS-PAGE analysis of bioconjugate formation. (**b**) SDS-PAGE analysis of type III AFP conjugated to a dendrimer. (Adapted with permission [5])

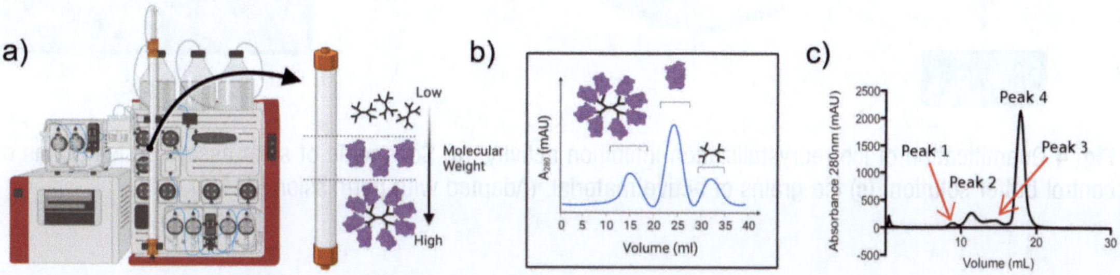

Fig. 3 Size-exclusion purification of protein–polymer bioconjugate. (**a**) Schematic workflow of size-exclusion purification. (**b**) Size-exclusion chromatogram of type III AFP–dendrimer bioconjugate. (Adapted with permission [5])

2. Prepare a sufficient volume of running buffer (~4 L of 50 mM Tris–HCl, 150 mM NaCl pH 7.4) for the entire run and connect it to the FPLC.
3. Wash the column with buffer, with at least two column volumes using appropriate flow rate depending on manufacturer's recommendation.
4. Start the chromatography run and load appropriate sample volume depending on column volume. Make sure to avoid air bubbles.
5. Collect separate fractions for the different peaks (*see* Fig. 3 and **Note 3**).
6. Compare the results of the FPLC analysis for the different samples to the controls run previously.

3.3 Activity Analysis of Conjugate

There are multiple ways to assess the various activities of IBPs and IBP–polymer conjugates; one way is to determine the ice recrystallization inhibition activity of the material by the SPLAT assay quantified by mean grain size. See Chap. 5 for thermal hysteresis measurement and Chap. 6 for additional IRI quantification method.

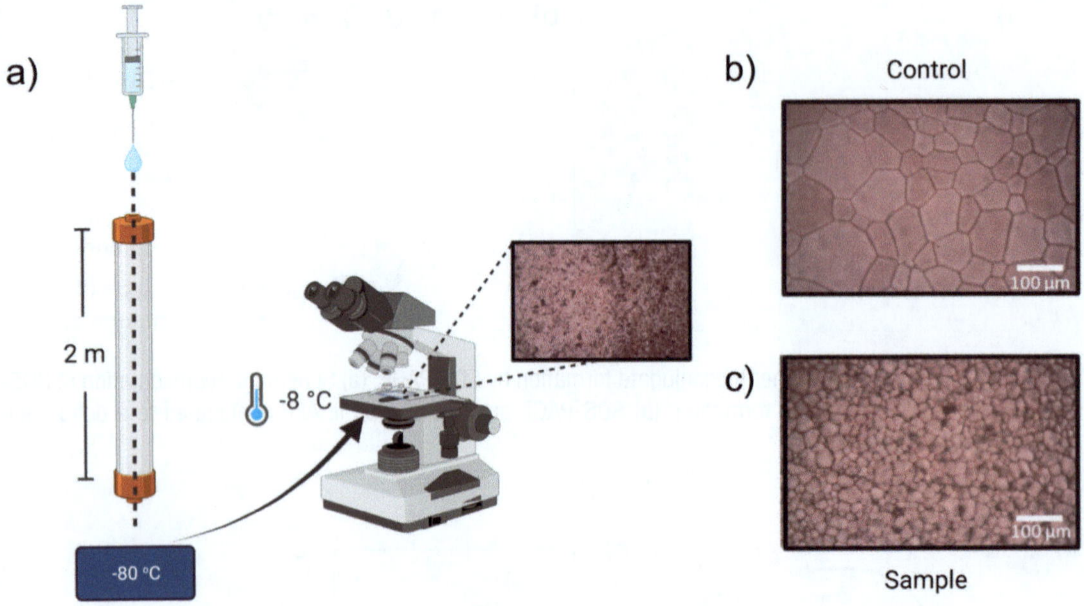

Fig. 4 Quantification of ice recrystallization inhibition activity. (**a**) Schematic of splat assay. (**b**) Ice grains of control buffer solution. (**c**) Ice grains of active material. (Adapted with permission [7, 12])

1. Cool a microscope slide atop an aluminum block surrounded by dry ice (approximately −80 °C).
2. Drop a 10 μL aliquot of the sample from approximately 2 m onto the chilled microscope slide (typically through an empty column) (*see* Fig. 4a).
3. Transfer the newly formed ice wafer to a cooling stage set to −8 °C paired with a microscope.
4. Incubate at −8 °C for 30 min.
5. Image and count the number of individual ice grains in the sample.
6. Calculate the mean grain size using the following formula:

$$\text{Mean grain size (MGS \%relative to control)} = \frac{\text{\# of crystals in control}}{\text{\# of crystals in sample}} \times 100$$

7. Use the appropriate control (i.e., buffer or inactive ice-binding protein).
8. Repeat at different concentrations and plot MGS versus concentration.

4 Notes

1. Representative SDS-PAGE of type III AFP conjugation reaction to the second-generation PAMAM dendrimer via SM $(PEG)_2$ heterobifunctional cross-linker is shown in Fig. 2. The shift in band size compared to control samples type III AFP, unmodified polymer, and the linker-modified polymer indicates successful cross-linking.
2. Prior to running the size-exclusion column, perform control runs on the following control samples on the FPLC to note when they will elute from the column: ice-binding protein, unmodified polymer, and the linker-modified polymer.
3. Representative FPLC chromatogram profile of type III AFP conjugation reaction to the second-generation PAMAM dendrimer via $SM(PEG)_2$ heterobifunctional cross-linker is shown in Fig. 3. Compare peaks to control chromatograms and run SDS-PAGE as described in 3.1 to further assess peak composition.
4. MALDI-TOF analysis of conjugate—Molecular weight determination of protein–polymer conjugates can be difficult due to polymer and conjugate polydispersity. Typically, matrix-assisted laser desorption/ionization time-of-flight (MALDI-TOF) mass spectrometry has been used for heterogeneous mixtures. The specific operating conditions will depend on the particular MADLDI-TOF instrument used. A representative MADLI-TOF profile for type III AFP conjugation reaction to the second-generation PAMAM dendrimer via $SM(PEG)_2$ heterobifunctional cross-linker is shown in Fig. 5. To interpret, calculate the specific mass of the polymer and the protein to be conjugated, and estimate how many proteins have been immobilized on the polymer.

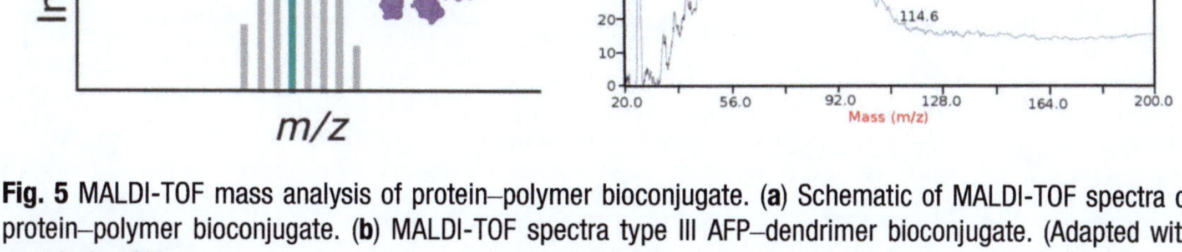

Fig. 5 MALDI-TOF mass analysis of protein–polymer bioconjugate. (**a**) Schematic of MALDI-TOF spectra of protein–polymer bioconjugate. (**b**) MALDI-TOF spectra type III AFP–dendrimer bioconjugate. (Adapted with permission [5])

References

1. Gauthier AM, Klok H-A (2008) Peptide/protein – polymer conjugates: synthetic strategies and design concepts. Chem Commun 0(23):2591–2611. https://doi.org/10.1039/B719689J
2. Grover GN, Maynard HD (2010) Protein–polymer conjugates: synthetic approaches by controlled radical polymerizations and interesting applications. Curr Opin Chem Biol 14(6): 818–827. https://doi.org/10.1016/j.cbpa.2010.10.008
3. Pelegri-O'Day EM, Lin E-W, Maynard HD (2014) Therapeutic protein–polymer conjugates: advancing beyond PEGylation. J Am Chem Soc 136(41):14323–14332. https://doi.org/10.1021/ja504390x
4. Davies PL (2014) Ice-binding proteins: a remarkable diversity of structures for stopping and starting ice growth. Trends Biochem Sci 39(11):548–555. https://doi.org/10.1016/j.tibs.2014.09.005
5. Stevens CA, Drori R, Zalis S, Braslavsky I, Davies PL (2015) Dendrimer-linked antifreeze proteins have superior activity and thermal recovery. Bioconjug Chem 26(9):1908–1915. https://doi.org/10.1021/acs.bioconjchem.5b00290
6. Esser-Kahn AP, Trang V, Francis MB (2010) Incorporation of antifreeze proteins into polymer coatings using site-selective bioconjugation. J Am Chem Soc 132(38):13264–13269. https://doi.org/10.1021/ja103038p
7. Wilkins LE, Hasan M, Fayter AER, Biggs C, Walker M, Gibson MI (2019) Site-specific conjugation of antifreeze proteins onto polymer-stabilized nanoparticles. Polym Chem 10(23):2986–2990. https://doi.org/10.1039/C8PY01719K
8. Hoon Ko J, Maynard D, H. (2018) A guide to maximizing the therapeutic potential of protein–polymer conjugates by rational design. Chem Soc Rev 47(24):8998–9014. https://doi.org/10.1039/C8CS00606G
9. Baker SL, Kaupbayeva B, Lathwal S, Das SR, Russell AJ, Matyjaszewski K (2019) Atom transfer radical polymerization for biorelated hybrid materials. Biomacromolecules 20(12): 4272–4298. https://doi.org/10.1021/acs.biomac.9b01271
10. Obermeyer AC, Olsen BD (2015) Synthesis and application of protein-containing block copolymers. ACS Macro Lett 4(1):101–110. https://doi.org/10.1021/mz500732e
11. Stevens CA, Semrau J, Chiriac D, Litschko M, Campbell RL, Langelaan DN, Smith SP, Davies PL, Allingham JS (2017) Peptide backbone circularization enhances antifreeze protein thermostability. Protein Sci 26(10): 1932–1941. https://doi.org/10.1002/pro.3228
12. Stevens CA, Bachtiger F, Kong X-D, Abriata LA, Sosso GC, Gibson MI, Klok H-A (2021) A minimalistic cyclic ice-binding peptide from phage display. Nat Commun 12(1):2675. https://doi.org/10.1038/s41467-021-22883-w

Chapter 16

Analysis of Ice-Binding Protein Evolution

Isaiah C. H. Box, Karin R. L. van der Burg, and Katie E. Marshall

Abstract

Discovering novel ice-binding proteins (IBPs) is important for understanding the evolution of IBPs but it is difficult to determine where resources should be directed in the search for novel IBPs. For this reason, we developed a simple bioinformatic approach for aiding in the determination of where to direct efforts in the search for IBPs. First, BLAST is used to obtain a candidate list of putative IBPs. Next, phylogenetic trees are constructed to map the candidate list of putative IBPs to determine if any patterns are forming. These candidate putative IBPs and their patterns are then assessed through the production of ancestral sequences and reverse BLAST searches, in addition to the use of IBP calculators, to determine which sequences should be cut to produce the final putative IBP list. Finally, we explain an avenue to investigate these putative IBPs further for the development of hypotheses on their evolution.

Key words Bioinformatics, BLAST, Phylogeny, Gene mapping

1 Introduction

While ice-binding proteins (IBPs) share their ability to bind to ice, they are diverse in structure [1, 2]. In fact, IBPs often bear more similarity to proteins that are unable to bind to ice than to other IBPs, with there being no universal ice-binding domain [1–3]. This makes the identification of IBPs near impossible without the use of laboratory measures of IBP activity such as thermal hysteresis, ice recrystallization inhibition, or ice nucleation [1, 2]. Performing these measures to determine the presence or absence of IBPs for each (sub) species that may benefit from these proteins is unrealistic. However, a bioinformatic approach to identifying putative IBPs could aid in directing laboratory efforts. This is important not only to expand on the list of known IBPs but also in researching the evolution of IBPs. A proper understanding of the distribution of IBPs across the various taxa must first be achieved before the evolution of IBPs across the tree of life can be fully contextualized. This simple protocol can be used to aid in accomplishing this by

Ran Drori and Corey A. Stevens (eds.), *Ice Binding Proteins: Methods and Protocols*, Methods in Molecular Biology, vol. 2730, https://doi.org/10.1007/978-1-0716-3503-2_16,

investigating IBPs at a habitat level [4] or down to a single species, and its applicability will continue to expand with the ever-growing list of genome-sequenced species.

2 Materials

1. Computer with a stable Internet connection.
2. Query list of NCBI accession codes for sequenced IBPs of interest (*see* **Note 1**) with confirmed ice-binding activity in FASTA format.
3. Search list of NCBI taxonomic ID codes for organisms of interest; this could be species specific or at the level of kingdoms/domains of life (*see* **Note 2**).
4. MEGA X [5] software (*see* **Note 3**).

3 Methods

The protocol below was performed using Windows 10 and a Chrome browser where applicable. Some adjustments may be required for previous versions of Windows, macOS, Linux, and/or other Internet browsers.

3.1 Search for Putative IBPs

Performing an initial search through the relevant molecular sequence data on NCBI allows for the production of a candidate list of putative IBPs. This candidate list of putative IBPs can then later be assessed to eliminate sequences unlikely to be IBPs to produce a final list of putative IBPs.

1. Access BLAST [6] through NCBI (URL: https://blast.ncbi.nlm.nih.gov/Blast.cgi) and select the BLAST [6] program that fits the query dataset and desired output. There are BLAST [6] programs that allow for searching nucleotide sequences against nucleotide sequences (blastn), amino acid sequences against amino acid sequences (blastp), or a combination of the two (blastx or tblastn; *see* **Note 4**).
2. Upload a FASTA file of the query list of IBPs in the "Enter Query Sequence" section of the BLAST [6] search page or enter the accession codes manually in the provided textbox.
3. In the "Organism" portion of the "Choose Search Set" section of the BLAST [6] search page, enter the taxonomic ID codes for the organisms of interest. This "Organism" search filter can also be used to exclude certain taxa, which will prevent overlap in the search organisms and the query list when performing searches at broader taxonomic levels.

4. Expand the "Algorithm Parameters" at the bottom of the BLAST [6] page and change the "Expect Threshold" to 1×10^{-5} (can be written as "1e-05"). This will increase the specificity of the search and is often seen as the default when using BLAST [6] as per Schäffer et al. [7]. Other values can be used, but 1×10^{-5} should be an upper limit.
5. All other BLAST [6] search parameters can be left at their default parameters unless otherwise desired (*see* **Note 5**). Use the blue question marks throughout the search page to help determine what changes may be necessary based on individual search goals.
6. Check the box called "Show results in new window" to allow adjustments to the BLAST [6] search without starting from scratch if there is something wrong with the output. Also, when performing the same search multiple times with slight variations, creating an NCBI account and saving the search strategy can save time going forward.
7. Once all BLAST [6] search settings are adjusted as desired, press the "BLAST" button to perform the search.
8. BLAST [6] will provide a search output with a list of sequences (if any; *see* **Note 6**) that match the search parameters, and from here, results can be downloaded as a FASTA file for later external analyses as well as other desired available file formats. To avoid downloading unwanted sequences, deselect them by clicking on the blue checkmark next to the sequence before downloading. Also, from the output page, a graphical summary of the results can be observed as well as sequence alignments of the queries and matching sequences from the search to help visualize the BLAST [6] search results (*see* Fig. 1).

3.2 Phylogenies

Phylogenetic trees can be used to group like sequences and allow for visualization of the similarity of putative IBPs. Different clustering of putative IBPs could provide evidence for different evolutionary hypotheses based on their taxonomic origin. For example, if IBPs from a single taxon are clustered, this could indicate that a particular IBP evolved in that taxon based on their shared ancestors. By contrast, if distantly related species cluster at a particular type of IBP in the phylogeny, it could indicate convergent evolution or even horizontal gene transfer.

1. Open the query IBP FASTA file in MEGA [5]. MEGA [5] will ask if the file is to be analyzed or aligned, select "Align," and this will open a window with the sequences of the queries. Sequences need to be aligned before a phylogeny can be produced from them.
2. Select the "W" along the top menu and then select the align prompt to align the sequences using the ClustalW algorithm

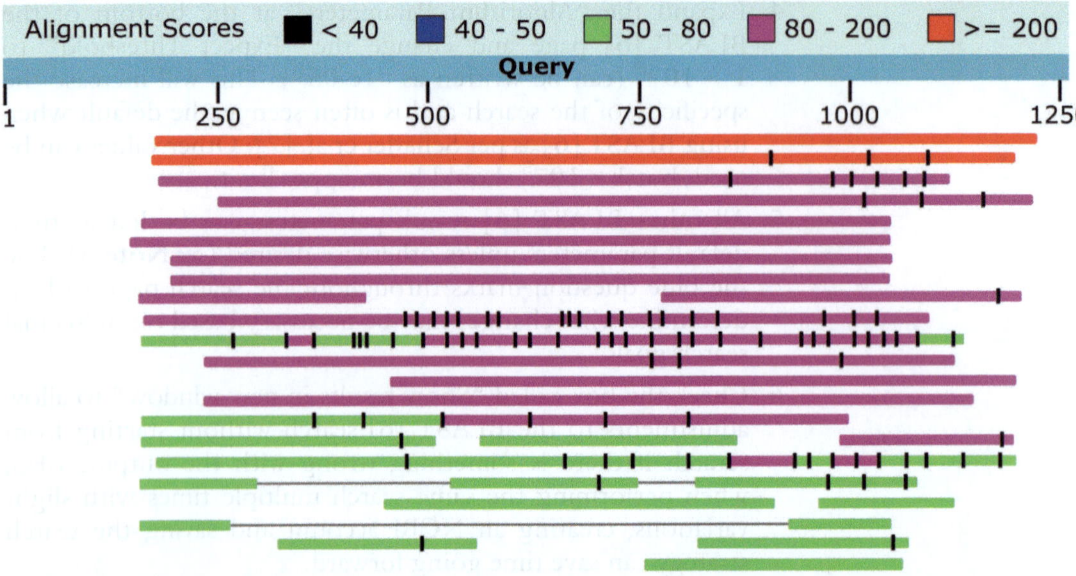

Fig. 1 Example of graphical alignment from BLAST [6] output of a bacterial ice-nucleating IBP query using the organism search list from Box et al. [4]. Black hash marks along the aligned sequences distinguish BLAST results along different regions of the same sequence. Numbers along the dashed "Query" line represent the number of amino acids

[8] (*see* **Note 7**). This will trigger a prompt asking to select all for the alignment; press "OK." Use the default alignment parameters unless the project requires otherwise.

3. Export the alignment as a MEGA [5] file using the "Data" tab at the top of the window; save it in a memorable folder with a recognizable name.
4. Leave the alignment window and return to the main MEGA [5] window, then select the "Models" tab, and then select "Find best DNA/protein models (ML)..." from the dropdown menu. Select the recently exported alignment MEGA [5] file and then leave the model selection at default settings and press "OK." This will determine the substitution model to use when constructing a phylogeny from the data. The model at the top of the resulting table is the one to be used going forward for that alignment.
5. Return to the main MEGA [5] window and select the "Phylogeny" tab followed by "Construct/test maximum likelihood tree..." from the dropdown menu and select the same exported alignment file. A window will pop up which is where the appropriate substitution model determined above can be selected. To improve results, select bootstrap method in the "Test of phylogeny" option, but this is not necessary as the IBP sequences being used should be quite disparate and the

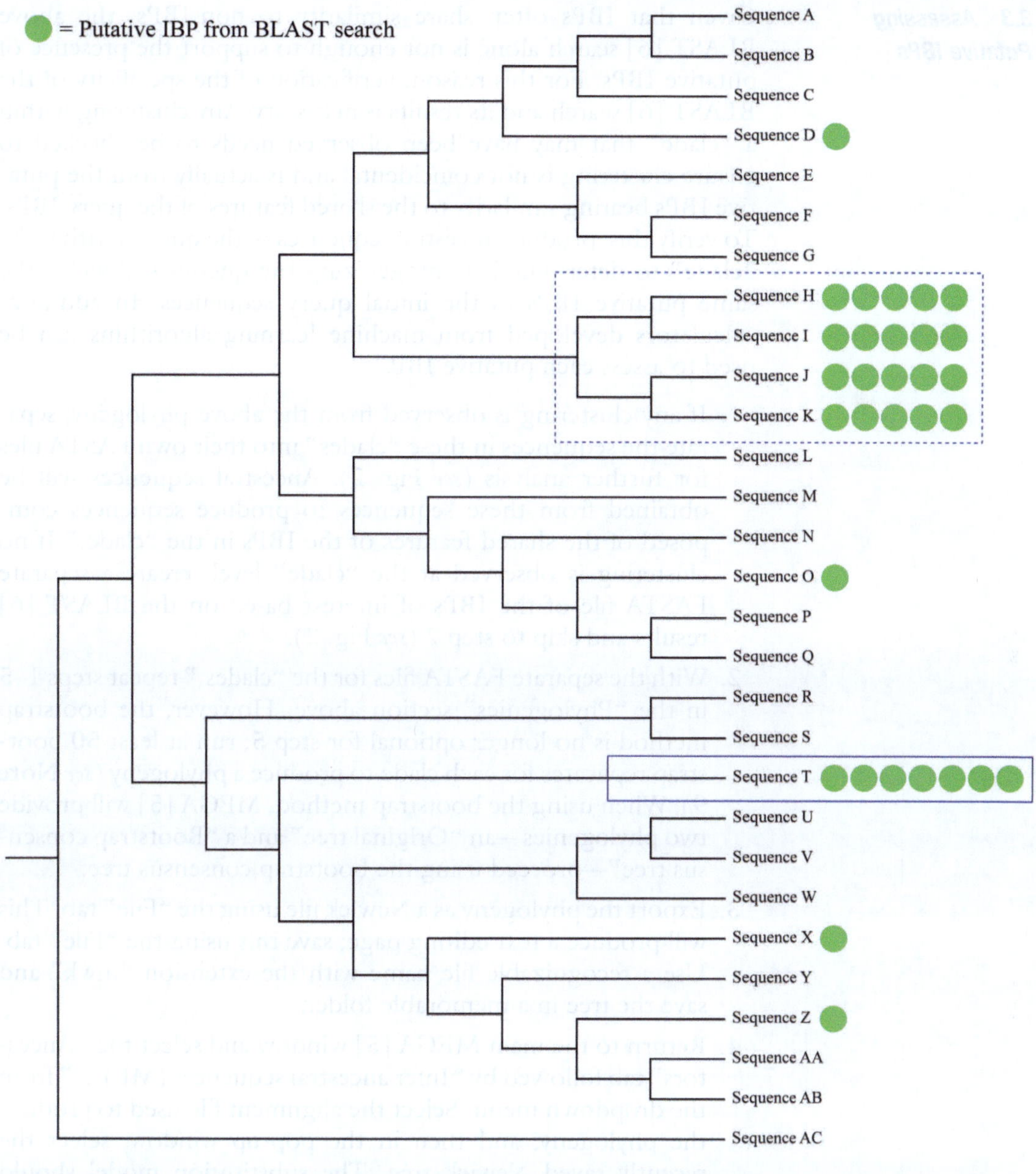

Fig. 2 Hypothetical query IBP phylogeny with putative IBPs from a BLAST [6] search mapped against it, demonstrating clustering at a "clade" (encompassed in the dashed lines). In addition, an example of clustering that does not form a "clade" but is still of interest is shown (encompassed in the solid lines)

goal of this phylogeny is to group similar sequences, not determine relatedness. Leave all other parameters at their default settings.

6. The resulting tree can be used to map BLAST [6] search results against it to visualize patterns that may be forming, such as any clustering of BLAST [6] results in certain IBP "clades" in the phylogeny (*see* Fig. 2 and **Note 8**).

3.3 Assessing Putative IBPs

Given that IBPs often share similarity to non-IBPs, the above BLAST [6] search alone is not enough to support the presence of putative IBPs. For this reason, verification of the specificity of the BLAST [6] search and its results is necessary. Any clustering within a "clade" that may have been observed needs to be checked to ensure clustering is not coincidental and is actually from the putative IBPs bearing similarity to the shared features of the query IBPs. To verify this, produce ancestral sequences of the queries within the "clade" to determine if homogenizing the queries still yields the same putative IBPs as the initial query sequences. In addition, calculators developed from machine learning algorithms can be used to assess each putative IBP.

1. If any clustering is observed from the above phylogeny, separate the sequences in these "clades" into their own FASTA files for further analysis (*see* Fig. 2). Ancestral sequences will be obtained from these sequences to produce sequences composed of the shared features of the IBPs in the "clade." If no clustering is observed at the "clade" level, create a separate FASTA file of the IBPs of interest based on the BLAST [6] results and skip to step 7 (*see* Fig. 2).
2. With the separate FASTA files for the "clades," repeat steps 1–5 in the "Phylogenies" section above. However, the bootstrap method is no longer optional for step 5; run at least 50 bootstrap replicates for each clade to produce a phylogeny (*see* **Note 9**). When using the bootstrap method, MEGA [5] will provide two phylogenies—an "Original tree" and a "Bootstrap consensus tree"—proceed using the bootstrap consensus tree.
3. Export the phylogeny as a Newick file using the "File" tab. This will produce a text editing page; save this using the "File" tab. Use a recognizable file name with the extension ".nwk" and save the tree in a memorable folder.
4. Return to the main MEGA [5] window and select the "Ancestors" tab followed by "Infer ancestral sequences (ML)..." from the dropdown menu. Select the alignment file used to produce the phylogeny, and then in the pop-up window, select the recently saved Newick tree. The substitution model should match the one used to produce the initial phylogeny; the rest can remain at their defaults (*see* **Note 10**). The resulting window will display a phylogeny resembling the one already produced, but this window has the calculated ancestral sequences hidden within it.
5. Open the "View" tab in this new window and then proceed to select "Show/hide" and then "Node IDs" in the resulting dropdown menus. This will show the numeric identifier for each node on the tree.

6. Open the "Ancestors" tab in the same window and select "Export most probable sequences" from the dropdown menu. This will produce an Excel file bearing the hypothesized ancestral sequences for each node in the phylogeny, identified by the numerals revealed in the previous step. These sequences can then be transferred to any text editing software and formatted as a FASTA file and saved with the extension ".FASTA" for future steps.
7. To verify the initial search for putative IBPs was specific, perform a BLAST [6] search against NCBI's non-redundant database (do not filter by organisms like in previous steps) using the ancestral sequences obtained above or IBP sequences of interest if skipping to this step. Verify that the top results are IBPs and not progenitor proteins or completely unrelated proteins (list of IBP progenitors can be found in Bildanova et al. [9]). If top results are non-IBP, then it is unlikely the BLAST [6] results from these sequences in the search organisms represent IBPs, and they should likely be omitted from future analyses.
8. Repeat the steps in the "Search for putative IBPs" section using the ancestral sequences obtained above to verify that the new sequences still yield similar BLAST [6] results with respect to the search organisms. BLAST [6] results yielded from the initial BLAST [6] search that are not repeated here are unlikely to be IBPs and should likely be omitted from future analyses.
9. Further assess the putative IBPs remaining by using IBP calculators (Table 1; *see* **Note 11**). No IBP calculator is reliable alone, and it is best to use at least two to get a consensus on which putative IBPs are most likely to be true IBPs [4]. To further assess how reliable the outputs of the calculators are, run your query IBPs through the calculators. From this, create a finalized putative IBP list (*see* **Note 12**).

3.4 Gene Mapping

Basic gene mapping of putative IBPs, in combination with any patterns seen in the distribution of putative IBPs in the above phylogeny, can provide insight into the potential evolution of the putative IBPs.

1. Select a putative IBP from the finalized list created above and locate the coding region for the putative IBP in the genome of the organism in NCBI by going to the nucleotide page and then selecting "Graphics" or "Show in genome data viewer" (*see* **Note 13**).
2. Note the coding regions surrounding that of the putative IBP and compile them (and their coinciding mRNA sequences separately if possible) into a FASTA file. Then record the size of the portion of the genome bearing all these coding regions using the nucleotide (nt) count found in the bottom corner of

Table 1
Examples of IBP calculators and how to access them for personal use

IBP calculator	Availability
AFP-Pred [3]	http://www3.ntu.edu.sg/home/EPNSugan/index_files/AFP-Pred.htm
AFP_PSSM [10]	No longer publicly available; contact authors
Target Freeze [11]	http://csbio.njust.edu.cn/bioinf/TargetFreeze
AFP-Ensemble [12]	No longer publicly available; contact authors
RAFP-Pred [13]	https://cloudstor.aarnet.edu.au/plus/index.php/s/Ye6gG8HncgatosK
iAFP-Ense [14]	No longer publicly available; contact authors
CryoProtect [15]	http://codes.bio/cryoprotect/
afpCOOL [16]	No longer publicly available; contact authors
AFP-LSE [17]	https://github.com/Shujaat123/AFP-LSE
Unnamed W-GDipc-LRMR-Ri Calculator [18]	https://github.com/Xia-xinnan/W-GDipc-LRMR-Ri

All calculators were developed from the IBP dataset of Kandaswamy et al. [3] or an adaptation of it (*see* **Note 1**)

the genome viewer and download a FASTA file of this entire portion of the genome as a "snapshot."

3. Based on available genome sequence data, select species related to the organism bearing the putative IBP of interest and access their genome pages on NCBI.
4. In these genome pages under "Related Information," select "Assembly" (*see* **Note 14**). From here, select "BLAST the assembly" (*see* **Note 15**).
5. Enter the coding regions and/or the mRNA sequences for the putative IBP and its surrounding genes in the query region and BLAST [6] the genomes.
6. Note the location of the coding region of the top results for each query sequence and determine if there is any synteny across any of the observed species and the species with the putative IBP.
7. Take a genome "snapshot" of matching size as above around the coding region for the top BLAST [6] result of the putative IBP in each of the related species.
8. Upload these "snapshot" FASTA files and the FASTA files for the coding regions or mRNA sequences into SimpleSynteny [19] (URL: https://www.dveltri.com/simplesynteny/index.html) to visualize the genomic synteny between species or lack thereof. From here, the formulation of hypotheses on how the putative IBP may have evolved can begin.

4 Notes

1. Examples of IBP lists that can be used for the query list can be found in literature [3, 4, 10, 12, 16] and can be expanded or shortened based on personal criteria. Note that each list of IBPs was constructed with different parameters and may include sequences not confirmed to have ice-binding ability or exclude confirmed ice-binding proteins.
2. The more molecular data available for the search organisms, the better. Organisms without at least a shotgun-sequenced genome will likely lack the molecular data sufficient for this protocol.
3. MEGA [5] requires a minimum 2048 MB of RAM and 500 MB of available hard disk space.
4. When using a BLAST [6] program that outputs nucleotide data, obtained sequences directly from the genome will need to be vetted for an open reading frame, to ensure the putative IBP can be expressed. This can be done through the Open Reading Frame Finder in NCBI (URL: https://www.ncbi.nlm.nih.gov/orffinder/).
5. The "Max Target Sequences" search parameter may need to be increased depending on the size of the search organism list.
6. The first sequence in the query list may not yield results from the BLAST [6] search; this does not mean none did. Select the dropdown menu called "Results for" and all sequences not marked with an "*" are sequences that yielded results.
7. Rather than ClustalW [8], the icon shaped like an arm can be selected to align using MUSCLE [20]. These two alignment algorithms differ in their approach, but for the purposes of this protocol, with disparate sequences being used, these differences are not consequential. However, if adapting this protocol for more identical sequences and search species, read up on these two algorithms to determine which best suits the data.
8. To instead map BLAST [6] results against a phylogeny of the search organisms, using the NCBI "Taxonomy Common Tree" tool in their "Taxonomy" menu (URL: https://www.ncbi.nlm.nih.gov/Taxonomy/CommonTree/wwwcmt.cgi) and entering the organisms of interest is a straightforward way to get a phylogeny of the search organisms. The final phylogeny can then be saved as a PHYLIP tree, which can be further edited for formatting in FigTree (http://tree.bio.ed.ac.uk/software/figtree/).
9. If the "clade" does not contain enough sequences, MEGA [5] will state that it is not possible to use the bootstrap method. In this case, it is fine to proceed without bootstrapping.

10. MEGA [5] may give an error saying that an outgroup is needed; press "OK." This will produce a pop-up window, which allows the appropriate sequence to be moved to the outgroup based on the phylogeny the ancestral sequences are being determined for.
11. The references provided are for IBP calculators that only require sequence data. There are IBP calculators that require 3D structure as well (e.g., Kozuch et al. [21]). Although the 3D structures of these putative IBPs are unknown, algorithms such as AlphaFold [22] are capable of producing them computationally.
12. To reiterate, none of these methods can be used to conclude that an IBP has been discovered. These simply allow for evidence to support the presence of IBPs in certain organisms. Conclusions can only be made through verification of expression of the protein(s) and verification of ice-binding activity by the protein(s) using laboratory methods.
13. Some genome projects are not at a sufficient stage of development to proceed. If the putative IBP is the only object within its respective contig when looking at the coding region, then this protocol cannot be followed further, and another protein must be selected.
14. Selecting the "Assembly" link may bring a page with more than one assembly. If this occurs, select the assembly that is at the highest assembly level (e.g., select chromosome over scaffold).
15. The genome pages have a link called "BLAST genome"; do not use these links. The BLAST [6] searches will not yield results.

Acknowledgments

Significant contributions to the development of this protocol were made by Ben Matthews. Trish Schulte and Jeff Richards also provided feedback that helped in the development of this protocol. KEM is funded by a NSERC Discovery Grant.

References

1. Davies PL (2014) Ice-binding proteins: a remarkable diversity of structures for stopping and starting ice growth. Trends Biochem Sci 39:548–555
2. Bar Dolev M, Braslavsky I, Davies PL (2016) Ice-binding proteins and their function. Annu Rev Biochem 85:515–542
3. Kandaswamy KK, Chou K-C, Martinetz T, Möller S, Suganthan PN, Sridharan S, Pugalenthi G (2011) AFP-Pred: a random forest approach for predicting antifreeze proteins from sequence-derived properties. J Theor Biol 270:56–62
4. Box ICH, Matthews BJ, Marshall KE (2022) Molecular evidence of intertidal habitats selecting for repeated ice-binding protein evolution in invertebrates. J Exp Biol 225:jeb243409
5. Kumar S, Stecher G, Li M, Knyaz C, Tamura K (2018) MEGA X: molecular evolutionary

genetics analysis across computing platforms. Mol Biol Evol 35:1547–1549
6. Altschul SF, Gish W, Miller W, Myers EW, Lipman DJ (1990) Basic local alignment search tool. J Mol Biol 215:403–410
7. Schäffer AA, Wolf YI, Ponting CP, Koonin EV, Aravind L, Altschul SF (1999) IMPALA: matching a protein sequence against a collection of PSI-BLAST-constructed position-specific score matrices. Bioinformatics 15: 1000–1011
8. Thompson JD, Higgins DG, Gibson TJ (1994) CLUSTAL W: improving the sensitivity of progressive multiple sequence alignment through sequence weighting, position-specific gap penalties and weight matrix choice. Nucleic Acids Res 22:4673–4680
9. Bildanova LL, Salina EA, Shumny VK (2012) Main properties and evolutionary features of antifreeze proteins. Russ J Genet Appl Res 3: 66–82
10. Zhao X, Ma Z, Yin M (2012) Using support vector machine and evolutionary profiles to predict antifreeze protein sequences. Int J Mol Sci 13:2196–2207
11. He X, Han K, Hu J, Yan H, Yang J-Y, Shen H-B, Yu D-J (2015) TargetFreeze: identifying antifreeze proteins via a combination of weights using sequence evolutionary information and pseudo amino acid composition. J Membr Biol 248:1005–1014
12. Yang R, Zhang C, Gao R, Zhang L (2015) An effective antifreeze protein predictor with ensemble classifiers and comprehensive sequence descriptors. Int J Mol Sci 16: 21191–21214
13. Khan S, Naseem I, Togneri R, Bennamoun M (2016) RAFP-Pred: robust prediction of antifreeze proteins using localized analysis of n-peptide compositions. IEEE/ACM Trans Comput Biol Bioinform 15:244–250
14. Xiao X, Hui M, Liu Z (2016) iAFP-Ense: an ensemble classifier for identifying antifreeze protein by incorporating grey model and PSSM into PseAAC. J Membr Biol 249:845–854
15. Pratiwi R, Malik AA, Schaduangrat N, Prachayasittikul V, Wikberg JES, Nantasenamat C, Shoombuatong W (2017) CryoProtect: a web server for classifying antifreeze proteins from nonantifreeze proteins. J Chem 2017:9861752
16. Eslami M, Zade RSH, Takalloo Z, Mahdevar G, Emamjomeh A, Sajedi RH, Zahiri J (2018) afpCOOL: a tool for antifreeze protein prediction. Heliyon 4:e00705
17. Usman M, Khan S, Lee JA (2020) AFP-LSE: antifreeze proteins prediction using latent space encoding of composition of k-spaced amino acid pairs. Sci Rep 10:7197
18. Wang S, Deng L, Xia X, Cao Z, Fei Y (2021) Predicting antifreeze proteins with weighted generalized dipeptide composition and multiregression feature selection ensemble. BMC Bioinform 22:340
19. Veltri D, Malapi-Wight M, Crouch JA (2016) SimpleSynteny: a web-based tool for visualization of microsynteny across multiple species. Nucleic Acids Res 44:W41–W45
20. Edgar RC (2004) MUSCLE: multiple sequence alignment with high accuracy and high throughput. Nucleic Acids Res 32:1792–1797
21. Kozuch DJ, Stillinger FH, Debenedetti PG (2018) Combined molecular dynamics and neural network method for predicting protein antifreeze activity. PNAS 115:13252–13257
22. Jumper J, Evans R, Pritzel A, Green T, Figurnov M, Ronneberger O et al (2021) Highly accurate protein structure prediction with AlphaFold. Nature 596:583–589

Index

Ran Drori and Corey Stevens (eds.), *Ice Binding Proteins: Methods and Protocols*, Methods in Molecular Biology, vol. 2730, https://doi.org/10.1007/978-1-0716-3503-2,

O

P

R

S

T

U

V

X

Y

MIX
Papier aus verantwortungsvollen Quellen
Paper from responsible sources
FSC® C105338

If you have any concerns about our products,
you can contact us on
ProductSafety@springernature.com

In case Publisher is established outside the EU,
the EU authorized representative is:
Springer Nature Customer Service Center GmbH
Europaplatz 3, 69115 Heidelberg, Germany

Printed by Libri Plureos GmbH
in Hamburg, Germany